ENVIRONMENTAL LABORATORY EXERCISES FOR INSTRUMENTAL ANALYSIS AND ENVIRONMENTAL CHEMISTRY

ENVIRONMENTAL LABORATORY EXERCISES FOR INSTRUMENTAL ANALYSIS AND ENVIRONMENTAL CHEMISTRY

FRANK M. DUNNIVANT
Whitman College

WILEY-INTERSCIENCE

A JOHN WILEY & SONS, INC., PUBLICATION

Library of Congress Cataloging-in-Publication Data:

Dunnivant, Frank M.
    Environmental laboratory exercises for instrumental analysis and environmental
chemistry / Frank M. Dunnivant
        p. cm.
    Includes index.
        ISBN 0-471-48856-9 (cloth)
    1. Environmental chemistry–Laboratory manuals.   2. Instrumental
        analysis–Laboratory manuals.   I. Title.
    TD193 .D86   2004
    628–dc22   2003023270

10 9 8 7 6 5 4 3 2 1

To my parents for nurturing
To my advisors for mentoring
To my students for questioning

CONTENTS

PREFACE   xi

ACKNOWLEDGMENTS   xiii

TO THE INSTRUCTOR   xv

PART 1   PRELIMINARY EXERCISES

1   How to Keep a Legally Defensible Laboratory Notebook   3

2   Statistical Analysis   7

3   Field Sampling Equipment for Environmental Samples   19

PART 2   EXPERIMENTS FOR AIR SAMPLES

4   Determination of Henry's Law Constants   33

5   Global Warming: Determining If a Gas Is Infrared Active   49

6   Monitoring the Presence of Hydrocarbons in Air around
    Gasoline Stations   61

PART 3   EXPERIMENTS FOR WATER SAMPLES

7   Determination of an Ion Balance for a Water Sample   73

8   Measuring the Concentration of Chlorinated Pesticides in
    Water Samples   83

9   Determination of Chloride, Bromide, and Fluoride in
    Water Samples   93

10 Analysis of Nickel Solutions by Ultraviolet–Visible
    Spectrometry   101

PART 4   EXPERIMENTS FOR HAZARDOUS WASTE

11 Determination of the Composition of Unleaded Gasoline
    Using Gas Chromatography   113

12 Precipitation of Metals from Hazardous Waste   123

13 Determination of the Nitroaromatics in Synthetic Wastewater
    from a Munitions Plant   143

14 Determination of a Surrogate Toxic Metal in a Simulated
    Hazardous Waste Sample   151

15 Reduction of Substituted Nitrobenzenes by Anaerobic
    Humic Acid Solutions   167

PART 5   EXPERIMENTS FOR SEDIMENT AND SOIL SAMPLES

16 Soxhlet Extraction and Analysis of a Soil or Sediment
    Sample Contaminated with n-Pentadecane   179

17 Determination of a Clay–Water Distribution Coefficient
    for Copper   191

PART 6   WET EXPERIMENTS

18 Determination of Dissolved Oxygen in Water Using the
    Winkler Method   207

19 Determination of the Biochemical Oxygen Demand of
    Sewage Influent   217

20 Determination of Inorganic and Organic Solids in Water Samples:
    Mass Balance Exercise   233

21 Determination of Alkalinity of Natural Waters   245

22 Determination of Hardness in a Water Sample   257

PART 7   FATE AND TRANSPORT CALCULATIONS

23 pC–pH Diagrams: Equilibrium Diagrams for Weak Acid and
    Base Systems   267

24   **Fate and Transport of Pollutants in Rivers and Streams**   277

25   **Fate and Transport of Pollutants in Lake Systems**   285

26   **Fate and Transport of Pollutants in Groundwater Systems**   293

27   **Transport of Pollutants in the Atmosphere**   303

28   **Biochemical Oxygen Demand and the Dissolved Oxygen Sag Curve in a Stream: Streeter–Phelps Equation**   317

APPENDIX A   Periodic Table   327

INDEX   329

PREFACE

My most vivid memory of my first professional job is the sheer horror and ineptitude that I felt when I was asked to analyze a hazardous waste sample for an analyte that had no standard protocol. Such was life in the early days of environmental monitoring, when chemists trained in the isolated walls of a laboratory were thrown into the real world of sediment, soil, and industrial waste samples. Today, chemists tend to be somewhat better prepared, but many still lack experience in developing procedures for problematic samples. My answer to this need for applied training is a book of laboratory experiments aimed at teaching upper-level undergraduate and graduate chemistry students how to analyze "dirty" samples. These experiments can be taught under the auspices of a standard instrumental analysis course or under more progressive courses, such as environmental chemistry or advanced analytical environmental techniques.

In preparing this book, I have kept in mind a number of chemical and analytical considerations, some steming from fundamental principles taught in every chemistry department, others specific to environmental chemistry. First, chemists planning to work in the environmental field need to be aware of the uncompromising need for explicit laboratory documentation. Chemistry departments start this life-long learning exercise in general chemistry, where we tell students that any classmate should be able to pick up his or her laboratory notebook and repeat the work. Environmental chemistry takes this training one step further in that the experiments and their documentation must also be completed in a manner that is legally defensible. By *legally defensible*, I mean ready to serve as courtroom evidence, as almost any laboratory monitoring, no matter how routine, can easily become evidence to prosecute an illegal polluter. Thus, laboratory notebooks must be maintained in a standardized format (subject to state or federal authorities and discipline); if they are not, cases may be dismissed. The introduction to this manual contains a list of commonly accepted documentation procedures. They are arranged so that instructors can select which level of documentation is suitable for their course.

A second feature of this manual is that it is designed to be a complete, stand-alone summary of a student's laboratory work. In the student version of the laboratory manual, each procedure contains background information, safety precautions, a list of chemicals and solutions needed, some data collection sheets, and a set of blank pages for the student to compile results and write a summary of findings. Thus, when each experiment is finished, students have a complete summary of their work that can be used as a laboratory portfolio during interviews at graduate schools or with potential employers.

A third theme, presented early in this book, is statistical analysis. Although many students entering environmental chemistry or instrumental analysis have briefly studied linear regression and Student's t test, a more rigorous treatment of these topics is needed in laboratories dealing with instrumentation. As I tell my students, few if any instrumental techniques yield absolute numbers; all instruments have to be calibrated to some extent, and the most common approach is a linear least squares regression. One of the first exercises that I conduct in my classes is to have students build a spreadsheet to perform linear least squares analysis and Student's t test. I have found that students understand data analysis techniques significantly better after this spreadsheet exercise, as opposed simply to quoting numbers from the regression of a calculator. An electronic copy of these spreadsheets (which I have students replicate) is included with the instructor's edition, and the spreadsheets can be used throughout the semester for a variety of instruments.

Fourth, the laboratory exercises in this manual are designed to teach environmental chemistry and instrumental analysis simultaneously. The experiments are organized by sample media into sections of air, water, hazardous waste, sediment/ soil, and wet techniques, and the manual includes a set of pollutant fate and transport simulation exercises, which are becoming more and more necessary in environmental chemistry courses. The laboratory experiments emphasize sampling, extraction, and instrumental analysis. Interactive software packages for pollutant fate and transport simulations, Fate and the pC-pH simulator, are included with the text.

Compiling the experiments for this manual has been a very educational experience for me, as I have reflected on which experiments work best in which setting. This information is given in the notes to the instructor. All of the experiments have been used in my courses, either environmental chemistry or instrumental analysis. More important for instructors using this manual, most experiments have a sample data set of the results expected, which is posted on the Wiley website. Each year I find these sample results most helpful in troubleshooting laboratories and identifying student mistakes.

FRANK M. DUNNIVANT
March 2004

ACKNOWLEDGMENTS

I would like to thank my reviewers, Samantha Saalfield of Whitman College, Dr. Cindy Lee of Clemson University, and Dr. John Ferry of the University of South Carolina. Their efforts have helped significantly in turning my original manuscript into a readable and useful document. I am indebted to the Whitman College students from my environmental chemistry and instrumental methods of analysis courses (2000–2003) for testing and debugging the procedures given in the manual and for supplying the typical student results given on the Wiley website. There are a number of software packages included with this manual that were created by Whitman College students and with funding from Whitman College and the National Collegiate Inventors and Innovators Alliance (NCIIA) program. I am especially indebted to Dan Danowski (Cornell University) and Josh Wnuk, Mark-Cody Reynolds, and Elliot Anders (all of Whitman College) for their programming efforts. Funding from the Dreyfus Foundation started our initial programming of EnviroLand, the previous version of Fate. Last, but not least, I am grateful the professors in the environmental engineering and science program at Clemson University for all of their efforts, training, and patience during my graduate degrees.

F.M.D.

TO THE INSTRUCTOR

This laboratory manual is designed for use courses in Instrumental Methods of Analysis and Environmental Chemistry. In fact, students from both of these courses were involved in the testing of these procedures. The procedures emphasize solution preparation, experimental setup, use of instrumentation, and evaluation of results. Given that not everyone is an environmental chemist, I have put together a list of experiments I use in instrumental analysis that are also used in environmental experiment. If you are unfamiliar with environmental chemistry I have included extensive background information on the environmental topic being studied and most chapters have a complete set of student data for your review (included in the on-line instructor's information). Indeed, one advantage of using this manual is that I have found students to be very interested in learning from an environmental viewpoint.

For instrumental analysis, of course, I use the experiments that emphasize the instruments a bit more than the solution preparation. There are certain exceptions to this statement, for example Chapter 14 (The Determination of a Surrogate Toxic Metal in a Simulated Hazardous Waste Sample), which stresses matrix effects and technique specificity (chelation, activity, or concentration). The following is the general plan I used for the course on Instrumental Methods of Analysis. It is based on two 3-hour laboratory periods each week.

Chapters 1 and 2 are given as introductory material but I usually have students build a spreadsheet for the statistics chapter.

UV-Vis spectroscopy	Chapter 10
Infrared spectroscopy	Chapter 5
Electrodes	Chapter 9 or 14

Atomic absorption or emission spectroscopy	Chapters 14 or 7
High performance liquid chromatography	Chapter 13
Gas chromatography	Chapters 6, 8, 11, or 16
Ion chromatography	Chapter 7
Mass spectrometry	any of the chromatography chapters

For environmental chemistry there are a variety of approaches. First, if you do not use this manual in a course in Instrumental Methods of Analysis you can select from all of the experiments. Second, if you use the approach given above for instrumental methods of analysis, there are still plenty of experiments left for use in environmental chemistry. I select from the following experiments.

Sampling	Chapter 2 (covered in lecture)
Mass balance, weighing and pipeting skills	Chapter 20
DO and BOD	Chapters 18 and 19
Global warming	Chapter 5
Environmental monitoring	Chapters 6, 8, 9, 13, 16, 21, or 22
Hazardous waste treatment	Chapter 12
Transformation reactions	Chapter 15
Distribution coefficients	Chapter 17
Chemical speciation	Chapter 23 (covered in lecture)
Pollutant fate and transport	Chapters 24 to 28 (covered in lecture)

An alternative is to design your environmental course completely around wet techniques.

Whichever way you choose to use this manual I hope that you will be satisfied with our efforts. We have done our best to provide student-tested procedures from an environmental perspective, detailed procedures for making solutions and unknown samples, example student data for troubleshooting and to supplement your students' experimental data, two user-friendly software packages (*The pC-pH Simulator*® and *Fate*®). Additionally, after you adopt the manual for use by your students you will have access to Wiley's on-line resources for this manual and you will be sent *The GC Tutorial and The HPLC Tutorial*. The downloadable instructor's manual can be obtained at `http://www.wiley.com/wileycda/ wileytitle/productcd-0471488569.html`. The latter two software packages are particularly helpful if students view them prior to attempting the chromatography experiments.

PART 1

PRELIMINARY EXERCISES

1

HOW TO KEEP A LEGALLY DEFENSIBLE LABORATORY NOTEBOOK

Proper recording of your laboratory data and upkeep of your laboratory notebook are essential to conducting good science. As your laboratory instructor will state, you should record sufficient detail in your notebook that another person of your skill level should be able to understand your procedures and comments and be able to reproduce all of your results. In government and industry (the real world), laboratory notebooks are legal documents. They can be used to apply for and defend patents, to show compliance or noncompliance with federal and state laws, and simply as record keeping. In the real world, lab notebooks start off as completely blank pages. You fill in all of your daily laboratory activities, including your conclusions. This laboratory manual is more organized than those used in the real world but will also serve as an example of your laboratory documentation, which will be an essential part of your future job. Except for a few cases, data collection sheets have been omitted intentionally because they are not always present in the real world. You should read the procedures carefully and understand them before you come to lab and have a data collection sheet ready in your laboratory notebook when you arrive in lab.

The laboratory notebook is the basis for your laboratory reports. The language you use in notebooks should be objective, factual, and free of your personal feelings, characterizations, speculation, or other terminology that is inappropriate. The notebook is *your* record of your or your group's work. Entries made by anyone other than the person to whom the notebook belongs must be dated and

Environmental Laboratory Exercises for Instrumental Analysis and Environmental Chemistry
By Frank M. Dunnivant
ISBN 0-471-48856-9   Copyright © 2004 John Wiley & Sons, Inc.

signed by the person making the entry. This may seem redundant since you will be dating and signing every page, but this is the standard policy used in government and industry.

Although you will quickly outgrow your laboratory notebook after graduation, you should realize that some laboratory notebooks are permanent records of a research project; that is, they are stored securely for years. The typical life of a laboratory notebook ranges from 10 to 25 years. Notebooks are also categorized by levels of use and include (1) a *working laboratory notebook* (one that is not yet complete and is currently being used to record information), (2) an *active laboratory notebook* (one that is complete but is needed as a reference to continue a project: for example, volume two of your notebook), and (3) an *inactive laboratory notebook* (one that is complete and no longer needed for quick reference).

The guidelines that follow have been collected from standard operating procedures (SOPs) of the U.S. Environmental Protection Agency and the U.S. Department of Energy as well as from my experience in a number of laboratory settings. These practices (and even more detailed ones) are also commonly used in industry. Your instructor will choose which guidelines are appropriate for your class and advise you to place a checkmark by those selected.

Your laboratory instructor will decide what heading or sections your data recording should be divided into, but these usually consist of a (1) a purpose statement, (2) prelaboratory instructions, (3) any modifications to the procedures assigned, (4) data collection, (5) interpretations, and (6) a brief summary of your conclusions. Although your laboratory reports will contain detailed interpretations and conclusions, you should include these in your laboratory notebook to provide a complete account of the laboratory exercise in your notebook. As you maintain your notebook, be aware that if you add simple notes, labels, or purpose statements throughout your data collection, it will make your account of the laboratory exercise much clearer a week later when you prepare your laboratory report.

Suggested Guidelines.   Check those that apply to your class.

☐ 1. Use this notebook for all original data, calculations, notes, and sketches.

☐ 2. Write all entries in indelible ink (non-water soluble).

☐ 3. The data collection sections are divided into separate experiments, and within each experiment all laboratory notebook entries should be in chronological order. Note that in the real world, you will maintain separate notebooks for each project you are working on. In your future employment, all entries will be made in chronological order and you will not be allowed to skip from page to page or leave any blank spaces.

☐ 4. Include a date and initials at the bottom of each page.

☐ 5. Make minor corrections by placing a single line through the entry and labeling it with your initials and the date.

☐ 6. Major alterations or changes to previous entries should appear as new entries, containing the current date and a cross-reference (page number) to the previous entries. In making your corrections, do not obscure or obliterate previous or incorrect entries.

☐ 7. Do not remove any pages from the laboratory notebook unless you are specifically advised to do so by your laboratory instructor.

☐ 8. If your laboratory manual does not include chart-holder pages, glue or otherwise securely fasten charts, drawings, and graphs in the area provided for each experiment.

☐ 9. Designate each blank unused page or portion of a page equal to or greater than one-fourth of a page with a diagonal line through the unused portion to indicate that portion of the page is intentionally being left blank. Along the line write "intentionally left blank," with your initials, and date it.

☐ 10. Reference to a name, catalog number, or instrument number should be made when nonstandard items are being used or when the laboratory contains more than one piece of that equipment.

2

STATISTICAL ANALYSIS

Purpose: One of the first lessons that you need to learn in instrumental analysis is that few, if any, instruments report direct measurements of concentration or activity without calibration of the instrument. Even laboratory balances need periodic calibration. More complicated instruments need even more involved calibration. Instruments respond to calibration standards in either a linear or an exponential manner, and exponential responses can easily be converted to a linear plot by log or natural log transformation. The goals of this first computer exercise are to create a linear least squares spreadsheet for analyzing calibration data and to learn to interpret the results of your spreadsheet. The goal of the second computer exercise is to create a spreadsheet for conducting a Student's t test for (1) comparing your results to a known reference standard, and (2) comparing two groups' results to each other. Student's t test helps you evaluate whether the results are acceptable. The final exercise in this computer laboratory is to review propagation of uncertainty calculations.

BACKGROUND

Today, most calculators can perform a linear least squares analysis, but the output from these calculators is limited. The spreadsheet you will create in this exercise will give error estimates for every parameter you estimate. Error estimates are very important in telling "how good" a result is. For example, if your estimate of the slope of a line is 2.34 and the standard deviation is plus or minus 4.23, the

Environmental Laboratory Exercises for Instrumental Analysis and Environmental Chemistry
By Frank M. Dunnivant
ISBN 0-471-48856-9   Copyright © 2004 John Wiley & Sons, Inc.

estimate is not very good. In addition, one of the most important parameters we will estimate with your spreadsheet is the standard deviation for your sample concentration. With your spreadsheet you will first conduct a linear least squares analysis for a calibration curve. Then we will use the unknown sample area, millivolts, or peak height to estimate the unknown sample concentration, and finally, we will calculate the standard deviation of your concentration estimate. This is one parameter that calculators do not typically estimate.

Equipment Needed

- Access to a computer lab or laptop computer
- A basic knowledge of spreadsheets
- Two computer disks or a zip disk for storing your work
- A calculator for checking your work

Programming Hints for Using Microsoft Excel

1. Formulas (calculations) must start with an "=".
2. The "$" locks a cell address when referencing cells in formulas, allowing you to lock rows, columns, or both.
3. Mathematical symbols are as you expect, except that "∧" represents a number used as an exponent.
4. Text is normally entered as text, but sometimes you may have to start a line with a single-quote symbol,'.

LINEAR LEAST SQUARES ANALYSIS

The first step in analyzing unknown samples is to have something (millivolts, peak area, peak height, absorbance, etc.) to reference to the instrument signal (instruments do not read concentration directly). To relate the signal to concentration, we create a calibration curve (line).

All of our calibration curves will be some form of linear relationship (line) of the form $y = mx + b$. We can relate signal to concentration with the equation

$$S = mc + S_{bl}$$

where S is the signal (absorbance, peak area, etc.) response, m the slope of the straight line, c the concentration of the analyte, and S_{bl} the instrumental signal (absorbance, etc.) for the blank. This is the calibration equation for a plot of the signal S on the y axis and C on the x axis. The signal (S_m) of the detection limit will be $S_m = S_{bl} + ks_{bl}$ (where $k = 3$). The detection limit (C_m) is an arrangement of $y = mx + b$, where $y = S_m$, m is the slope, b is the y intercept, and x is the minimum concentration or detection limit.

We will usually collect a set of data correlating S to c. Examples of S include (1) light absorbance in spectroscopy, (2) peak height in chromatography, or (3) peak area in chromatography. We will plot our data set on linear graph paper or using a spreadsheet and develop an equation for the line connecting the data points. We define the difference between the point on the line and the measured data point as the residual (in the x and y directions).

For calculation purposes we use the following equations (S's are the sum of squared error or residuals):

$$S_{xx} = \sum (x_i - \bar{x})^2 = \sum (x_i^2) - \frac{(\sum x_i)^2}{N}$$

$$S_{yy} = \sum (y_i - \bar{y})^2 = \sum (y_i^2) - \frac{(\sum y_i)^2}{N}$$

$$S_{xy} = \sum (x_i - \bar{x})(y_i - \bar{y}) = \sum x_i y_i - \frac{\sum x_i \sum y_i}{N}$$

where x_i and y_i are individual observations, N is the number of data pairs, and $\bar{x}$ and $\bar{y}$ are the average values of the observations. Six useful quantities can be computed from these.

1. The slope of the line (m) is $m = S_{xy}/S_{xx}$.
2. The y intercept (b) is $b = y - mx$.
3. The standard deviation of the residuals (s_y) is given by

$$s_y = \sqrt{\frac{S_{yy} - m^2 S_{xx}}{N - 2}}$$

4. The standard deviation of the slope is

$$s_m = \frac{s_y}{\sqrt{S_{xx}}}$$

5. The standard deviation of the intercept (s_b) is

$$s_b = s_y \sqrt{\frac{\sum (x_i^2)}{N \sum (x_i^2) - (\sum x_i)^2}} = s_y \sqrt{\frac{1}{N - (\sum x_i)^2 / \sum (x_i^2)}}$$

6. The standard deviation for analytical results obtained with the calibration curve (s_c) is

$$s_c = \frac{s_y}{m} \sqrt{\frac{1}{L} + \frac{1}{N} + \frac{(\bar{y}_c - \bar{y})^2}{m^2 S_{xx}}}$$

where $\bar{y}_c$ is the mean signal value for the unknown sample, L the number of times the sample is analyzed, N the number of standards in the calibration curve, and $\bar{y}$ is the mean signal value of the y calibration observations (from standards). Thus, the final result will be a value (the analytical result) plus or minus another value (the standard deviation, s_c).

It is important to note what s_c refers to—it is the error of your sample concentration according to the linear least squares analysis. Since the equation for s_c in case 6 does not account for any error or deviation in your sample replicates (due to either sample preparation error such as pipetting or concentration variations in your sampling technique), s_c does not account for all sources of error in precision. To account for the latter errors, you need to make a standard deviation calculation on your sample replicates. The sequence of dilutions and other factors can be accounted for in a propagation of uncertainty (covered at the end of the chapter).

Most calculators have an r or r^2 key and you may know that the closer this value is to 1.00, the better. This number comes from

$$r = \frac{\sum x_i y_i}{\sqrt{\sum (x_i^2) \sum (y_i^2)}}$$

r (and r^2) is called the *coefficient of regression* or *regression coefficient*.

Table 2-1 is the printout of a spreadsheet using the equations described above. Note that only the numbers in boldface type are entry numbers (entered directly rather than calculated); all other cells contain equations for calculating the given parameters. This spreadsheet can be used in all of the exercises in this manual for analyzing your instrument calibration data. The data in Table 2-1 were obtained from students measuring magnesium on a flame atomic absorption spectrometer.

STUDENT'S t TEST

After you obtain an average value for a sample, you will want to know if it is within an acceptable range of the true value, or you may want to compare mean values obtained from two different techniques. We can do this with a statistical technique called *Student's t test*. To perform this test, we simply rearrange the equation for the confidence limits to

$$\bar{x} - \mu = \pm \frac{t \cdot \text{s.d.}}{\sqrt{N}} \tag{2-1}$$

where $\bar{x}$ is the mean of your measurements, μ the known or true value of the sample, t the value from the t table, s.d. the standard deviation, and N the number of replicates that you analyzed.

In the first application of the t test, we are basically looking at the acceptable difference between the measured value and the true value. The overall comparison

TABLE 2-1. Spreadsheet for Conducting a Linear Least Squares Regression Analysis

Template
Mg data from last year
Linear Least Squares Analysis
Number of Standard Observations = 6
Replicates for Sample Unknowns= 5
Units of Standard= ppm

Number	x-Value Conc.	y-Value Signal	xy	x-squared	y-squared
1	0.5	0.005	0.0025	0.25	0.000025
2	1	0.012	0.012	1	0.000144
3	2	0.027	0.054	4	0.000729
4	5	0.067	0.335	25	0.004489
5	10	0.122	1.22	100	0.014884
6	20	0.238	4.76	400	0.056644
7			0	0	0
8			0	0	0
9			0	0	0
10			0	0	0
11			0	0	0
Sums	38.5	0.471	6.3835	530.25	0.076915
Means	6.416667	0.0785			

$S_{xx}=$ 283 Sum of Squared Error for the mean of x
$S_{yy}=$ 0.0399 Sum of Squared Error for the mean of y
$S_{xy}=$ 3.36 Sum of Squared Error for x and y
$m=$ 0.0119 Slope
$b=$ 0.00234 y-intercept
$s_y=$ 0.00349 Standard Deviation of the Residuals
$s_m=$ 0.000207 Standard Deviation of the Slope
$s_b=$ 0.00195 Standard Deviation of the Intercept
$r=$ 1.000
$r^2=$ 0.999

Calc. of Minimum Detection Limit

Blanks Signal	$(x-\bar{x})^2$ 5 replicates
-0.001	4E-08
-0.002	6.4E-07
-0.001	4E-08
0.000	1.44E-06
-0.002	6.4E-07

mean of blanks= -0.001
std. dev. of blks = 0.000837

$S(m) = mean(S(blk)) + 3 s(blk)$
$S(m) =$ 0.00131
$c(m) =$ 0.211 Minimum Detection Limit

Unknowns	Signal Rep 1	Signal Rep 2	Signal Rep 3	Signal Rep 4	Signal Rep 5	Conc 1	Conc 2	Conc 3	Conc 4	Conc 5	Mean Abs.	Mean ppm	s_c
cold water	0.062	0.063	0.06	0.062	0.06	5.03	5.11	4.86	5.03	4.98	0.061	4.98	0.18
hot water	0.063	0.062	0.059	0.061	0.062	5.11	5.03	4.77	4.94	5.03	0.061	4.98	0.18
city tap	0.019	0.020	0.018	0.018	0.018	1.40	1.49	1.32	1.32	1.32	0.019	1.37	0.20
pond water	0.025	0.024	0.023	0.023	0.024	1.91	1.82	1.74	1.74	1.82	0.024	1.81	0.20
lab sample	0.064	0.066	0.064	0.064	0.063	5.19	5.36	5.19	5.19	5.11	0.064	5.21	0.18

s_c = std. dev. of an unknown

TABLE 2-2. Student's *t* Test of Sample Data

Statistics for Replicate Analyses

DATA SET 1

Units for Observation=	Abs	
Number of Replicates=	5	
DF	4	
alpha	0.05	
t Value from Table =	2.776451	=TINV(alpha,DF)

Replicate	Observation ppm Mg	(x-mean)	(x-mean)2
1	5.20	-0.0168	0.0002822
2	5.36	0.1512	0.0228614
3	5.20	-0.0168	0.0002822
4	5.20	-0.0168	0.0002822
5	5.11	-0.1008	0.0101606
6			
7			
8			
9			
10			
Sum	26.059		0.0338688

Mean =	5.21	
s =	0.0920	Standard Deviation of the Mean
CV (%) =	1.77	Coefficient of Variation
CL (+ or -)	0.114	
upper CL	5.33	Upper Confidence Limits on the Mean
lower CL	5.10	Lower Confidence Limits on the Mean

Say that the true value is = 5 ppm

Exp mean - true value =	0.212	
Calc Difference from Eqn=	0.1142549	Eqn

If (Exp mean - true mean) is greater than (Calc Difference from Eqn)
then the presence of bias is suggested
If (Calc Difference from Eqn) is greater than (Exp mean - true mean)
then no bias has been demonstrated
Is bias present?   YES   =IF(ABS(+D32)>D33,"YES","NO")

DATA SET 2

Units for Observation=	Abs	
Number of Replicates=	5	
DF	4	
alpha	0.05	
t Value from Table =	2.77645	=TINV(alpha,DF)

Replicate	Observation ppm Mg	(x-mean)	(x-mean)2
1	9.77	0.0928	0.008612
2	9.66	-0.0232	0.000538
3	9.54	-0.1392	0.019377
4	9.66	-0.0232	0.000538
5	9.77	0.0928	0.008612
6			
7			
8			
9			
10			
Sum	48.406		0.037677

Mean =	9.6812	
s =	0.0971	Standard Deviation of the Mean
CV (%) =	1.002	Coefficient of Variation
CL (+ or -)	0.1205	
upper CL	9.802	Upper Confidence Limits on the Mean
lower CL	9.561	Lower Confidence Limits on the Mean

Say that the true value is = 5 ppm

Exp mean - true value =	4.6812	
Calc Difference from Eqn=	0.120507	Eqn

If (Exp mean - true mean) is greater than (Calc Difference from Eqn)
then the presence of bias is suggested
If (Calc Difference from Eqn) is greater than (Exp mean - true mean)
then no bias has been demonstrated
Is bias present?   YES   =IF(ABS(+D32)>D33,"YES","NO")

POOLED DATA: COMPARING MEANS

DF	8	
alpha	0.05	DF = N(1) + N(2) - 2
t Value from Table =	2.306006	=TINV(alpha,DF)
s(pooled) =	0.0945685	Observed
Mean(1) -Mean(2) =	-4.4694	Observed
Eqn x(1) - x(2) = (+ or -)	0.137923	Difference in means from equation (t*s)/n's

If (Obs mean(1) - mean(2)) is greater than (Calc Difference from Eqn)
then the presence of bias is suggested
If (Calc Difference mean from Eqn) is greater than (Obs mean(1) - mean(2))
then no bias has been demonstrated
Is bias present?   YES   IF(ABS(+D45)>D46,"YES","NO")

is based on consideration of a t value, the standard deviation, and the number of observations. The t values are taken from tables such as the those in a quantitative analysis or instrumental analysis textbook, and you must pick a confidence interval and the degrees of freedom (this will be $N - 1$ for this test). If the experimental (observed) value of $x - \mu$ is larger than the value of $x - \mu$ calculated from the right side of equation (2-1), the presence of *bias* in the method is suggested; in other words, the experimental and true values are statistically different. If, on the other hand, the value calculated by the right side of the equation is larger, no bias has been demonstrated.

A more useful but difficult procedure can be performed to compare the mean results from two experiments or techniques. This uses the following equation:

$$\bar{x}_1 - \bar{x}_2 = \pm \frac{t \cdot \text{s.d.}_{\text{pooled}}}{\sqrt{n_1 n_2 / (n_1 + n_2)}}$$

$$s_{\text{pooled}} = \sqrt{\frac{s_1^2 (n_1 - 1) + s_2^2 (n_2 - 1)}{n_1 + n_2 - 2}} \tag{2-2}$$

where s_1 and s_2 are the respective standard deviations about each mean and n_1 and n_2 are the number of observations in each mean. In this case the degrees of freedom in the t table will be $N - 2$ (2 because you are using two s^2 values). As in the procedure above, if the experimental (observed) value of $x_1 - x_2$ is larger than the value of $x_1 - x_2$ calculated from equation (2-2), there is a basis for saying that the two techniques are different. If, on the other hand, the value calculated by the equation is larger, no basis is present for saying that the two techniques are different (i.e., the value from the equation gives the tolerance or level of acceptable error). Also note that if you use the 95% CI, your result will include 95 out of 100 analytical results and that 5 of the 100 will fall outside the range.

Table 2-2 conducts both of the t tests mentioned above and will serve as your template for creating your own spreadsheet. Again, numbers in boldface type are the only numbers that you will change when using this spreadsheet. The other cells contain equations for calculating each parameter estimate.

PROPAGATION OF UNCERTAINTY

The linear least squares analysis provides a way of predicting a concentration value for an unknown sample and provides error estimates, in the form of standard deviations, for each estimated parameter. However, the final calculation that you made in the spreadsheet, s_c, only incorporates error associated with the linear least squares regression. An equally important value is the propagation of uncertainty (POU) resulting from multiple dilutions and weighing events. Tables 2-3 to 2-6 show the tolerances of balances and class A glassware that are used in the POU analysis. POU equations for each type of mathematical function are shown in Table 2-7.

TABLE 2-3. Tolerances for Laboratory Balance Weights

Denomination (g)	Tolerance (mg)		Denomination (mg)	Tolerance (mg)	
	Class 1	Class 2		Class 1	Class 2
500	1.2	2.5	500	0.010	0.025
200	0.50	1.0	200	0.010	0.025
100	0.25	0.50	100	0.010	0.025
50	0.12	0.25	50	0.010	0.014
20	0.074	0.10	20	0.010	0.014
10	0.050	0.074	10	0.010	0.014
5	0.034	0.054	5	0.010	0.014
2	0.034	0.054	2	0.010	0.014
1	0.034	0.054	1	0.010	0.014

Source: Harris (1999).

TABLE 2-4. Tolerances of Class A Burets

Buret Volume (mL)	Smallest Graduation (mL)	Tolerance (mL)
5	0.01	±0.01
10	0.05 or 0.02	±0.02
25	0.1	±0.03
50	0.1	±0.05
100	0.2	±0.10

Source: Harris (1999).

TABLE 2-5. Tolerances of Class A Volumetric Flasks

Flask Capacity (mL)	Tolerance (mL)	Flask Capacity (mL)	Tolerance (mL)
1	±0.02	100	±0.08
2	±0.02	200	±0.10
5	±0.02	250	±0.12
10	±0.02	500	±0.20
25	±0.03	1000	±0.30
50	±0.05	2000	±0.50

Source: Harris (1999).

TABLE 2-6. Tolerances of Class A Transfer Pipets (Harris, 1999)

Volume (mL)	Tolerance (mL)	Volume (mL)	Tolerance (mL)
0.5	±0.006	10	±0.02
1	±0.006	15	±0.03
2	±0.006	20	±0.03
3	±0.01	25	±0.03
4	±0.01	50	±0.05
5	±0.01	100	±0.08

Source: Harris (1999).

TABLE 2-7. Error Propagation in Arithmetic Calculations

Type of Calculation	Example	Standard Deviation of x
Addition or subtraction	$x = p + q - r$	$s_x = s_p^2 + s_q^2 + s_r^2$
Multiplication or division	$x = p(q/r)$	$\dfrac{s_x}{x} = \sqrt{\left(\dfrac{s_p}{p}\right)^2 + \left(\dfrac{s_q}{q}\right)^2 + \left(\dfrac{s_r}{r}\right)^2}$
Exponentiation	$x = p^y$	$\dfrac{s_x}{x} = y\dfrac{s_p}{p}$
Logarithm	$x = \log_{10}p$	$s_x = 0.434\dfrac{s_p}{p}$
Antilogarithm	$X = \text{antilog}_{10}p$	$\dfrac{s_x}{x} = 2.303 s_p$

Source: Skoog et al. (1998).

The use of these and other tolerances is illustrated in the following example. We weigh out 10.00 g of sample, extract it into 100 mL of solvent, make a 1 : 10 dilution, inject 1.0 µL into a GC, and calculate the concentration.

Operation	Raw Value of Operation	Error Associated with Each Operation (as $\pm s$)
Weighing	10.00 g	0.05
Extraction efficiency	0.95	0.02
Extraction volume	100.00 mL	0.02
Dilution 1	10.00	0.01
Injection volume	1.00×10^{-6} L	0.01×10^{-6}
Calculation of concentration (from linear least squares analysis)	1.14 pg/µL	0.05

Concentration of compound in (µg compound/g of sample)

$$= \dfrac{\overset{\text{conversion factor}}{\underset{(\mu g/10^6\,pg)}{}} \times \overset{(\text{peak Area} - b)/m}{\underset{(1.14\,pg/1\,\mu L)}{}} \times \overset{\text{solvent vol.}}{\underset{(100{,}000\,\mu L)}{}} \times \overset{\text{dil. 1}}{\underset{(10\,mL/1mL)}{}}}{\underset{\text{weight}}{10.00\,g} \times \underset{\text{ext. eff.}}{0.95}}$$

$= 0.120\ \mu g/g$

We use the standard deviation associated with each measurement to calculate the propagation of uncertainty (equations are shown in Table 2-7; in this case we use the example for multiplication but note that some of these may already have been calculated using addition or exponential error equations):

$$\frac{s_x}{x} = \sqrt{\left(\frac{0.05}{1.14}\right)^2 + \left(\frac{0.01 \times 10^{-6}}{1.00 \times 10^{-6}}\right)^2 + \left(\frac{0.01}{10.00}\right)^2 + \left(\frac{0.020}{0.95}\right)^2 + \left(\frac{0.05}{10.00}\right)^2}$$

$$= \sqrt{\underset{\text{calc.}}{0.00192} + \underset{\text{injection}}{0.00010} + \underset{\text{dil. 1}}{1.0 \times 10^{-6}} + \underset{\text{ext. eff.}}{0.00044} + \underset{\text{weight}}{2.5 \times 10^{-5}}}$$

Note that by comparing various errors, you can see which step in your procedure contributes the most error. In this case it is the calculation from the linear least squares analysis that commonly contributes most error to the standard deviation of the sample:

$$\frac{s_x}{x} = \pm\sqrt{0.00249} = \pm 0.0498$$

$$\text{absolute error} = \frac{s_x}{x}x = (\pm 0.0498)\,(0.120\,\mu g/g) = \pm 0.00598$$

Thus, the answer you report (with complete error) should be $0.120\,\mu g/g \pm 0.006$.

REFERENCES

Harris, D. C., *Quantitative Chemical Analysis*, 5th ed., W.H. Freeman, New York, 1999.

Skoog, D. A., F. J. Holler, and T. A. Nieman, *Principles of Instrumental Analysis*, 5th ed., Harcourt Brace College Publishing, Philadelphia, 1998.

ASSIGNMENT

1. Your first task is to create two spreadsheets that look identical to the ones in Tables 2-1 and 2-2. (Your instructor may choose to give you these on a disk to save time so that you can spend more time developing your analytical technique in the laboratory.) During the first laboratory period, you will create a linear least squares analysis sheet. For the second laboratory period you will create a spreadsheet for conducting a Student's t test. When you actually use the spreadsheet for calibrating an instrument, data should only be entered into cells containing boldface numbers; all other cells should contain equations that will not be changed (and can be locked to ensure that these cells do not change).

2. Next, calculate the propagation of uncertainty for the following set of data. Most quantitative measurements require several steps in a given procedure, including weighing, dilution, and various quantification approaches. Each of these processes has an associated error. Suppose that you are analyzing a liver sample for a given toxin X. You weigh 1 g of liver, dry it, extract it, and analyze your dilution. The steps, and the error associated with each step, are summarized in the following outline.

Operation	Value of Operation	Error Associated with Each Operation (as $\pm s$)
Weight (of wet liver) (g)	1.05	0.01
Determination of dry weight (g dry liver/g wet liver)	0.40	0.05
Total volume that toxin is extracted into	100 mL	0.05 mL
Extraction efficiency	0.90	0.05
Extraction volume	10. mL	0.01
Volume of solvent analyzed	1.00 μL	0.05 μL
Error from least squares analysis and calibration curve (the amount detected in 1.00 μL of injected solvent)	5.62 pg	0.08 pg

Calculate the concentration of toxin X in your original sample (in μg/g on a dry liver basis) and the total error associated with the measurement (propagation of error). Report concentrations in micrograms of toxin per gram of dry liver. Show all calculations for credit.

What do you turn in?

1. Supply a one-page printout (adjusted to fit onto one page) of each spreadsheet.

2. Before you turn in your spreadsheets, change the format of all column data to show three or four significant figures (whichever is correct).

3. Explain your linear least squares analysis and Student's t-test results (approximately one page each, typed).

Here are some things to include in your write-up. Basically, you should give an intelligent, statistically sound discussion of your data. Give:

- The equation of the line
- The signal-to-noise ratios for your analysis
- The minimum detection limit

Consider the following questions:

- Was bias indicated in your analysis of the unknown (the 5-ppm sample) and the true value?
- Were the results from the two groups comparable?
- How do the numbers compare to the results from your calculator?
- What shortcomings does your calculator have (if any)?

DATA COLLECTION SHEET

DATA COLLECTION SHEET

3

FIELD SAMPLING EQUIPMENT FOR ENVIRONMENTAL SAMPLES

BACKGROUND

The first and in many cases the most important step in any environmental monitoring plan is sampling. This may seem like an easy part of the process, but if a representative sample of a site is not taken properly, results obtained from analyzing the sample on a $100,000 instrument will be worthless. A bad sample can result from taking a sample at an inappropriate location, not taking the sample properly, not preserving the sample properly, or storing the sample too long. Many of these problems will not concern you directly today because most governmental and nongovernmental agencies and industries have developed clear sampling and analysis plans (SAPs). These will be stated clearly in the standard operating procedures (SOPs) where you work, so it would be pointless to teach you one set of procedures without knowing where you will be working in the future. There-fore, the purpose of this chapter is to introduce you to some of the standard sampling equipment used in environmental sampling. We divide the areas into atmospheric, surface water, groundwater, sediment/sludge, and soil samples, although many of these techniques are also relevant to hazardous waste.

It should be noted that most of the sampling equipment can be made of plastic, Teflon, or stainless steel, depending on your analyte. For example, plastic is generally used when analyzing metals, whereas stainless steel or Teflon is used when analyzing for organic compounds. Many of the sampling tools shown in the figures can be custom-made of specific materials.

Environmental Laboratory Exercises for Instrumental Analysis and Environmental Chemistry
By Frank M. Dunnivant
ISBN 0-471-48856-9   Copyright © 2004 John Wiley & Sons, Inc.

ATMOSPHERIC SAMPLING

Water samples (rain, snow, and ice) can be obtained using a sampling system as simple as a plastic or stainless steel bucket or as sophisticated as the automated sampler shown in Figure 3-1. Other types of atmospheric samplers actually have sensors to detect if it is precipitating or sunny and take wet or dry (particulate) samples. For sampling in remote areas, solar-powered units are available (Figure 3-2). Strictly dry particulate samples can be obtained using a high-volume atmospheric sampler like the one shown in Figure 3-3. Air enters the unit at the top and is pulled through a large weighed filter (typically, the size of a 8.5 by 11-inch piece of notepaper). The mesh or pore size of the filter paper can be selected to collect a specific particle size. This approach allows for the total mass of particles to be determined as well as for laboratory analysis of the particles.

Sampling indoor and outdoor gases is relatively easy using a portable personnel pump like the one shown in Figure 3-4. In this system the flow rate of the pump is calibrated to a specified value (typically, 2.0 L/min). A sampling tube containing a resin that is designed specifically to sample a compound or set of compounds is attached to the pump. The pump is actually a vacuum pump that pulls air first through the sample collection tube and then into the pump, thus not allowing the pumping system to contaminate the air. The resin tubes are returned

Figure 3-1. Model 200 wet-only rainwater sampler designed by Ecotech Pty Ltd, Blackburn, Victoria. (Reproduced with permission from Ecotech Pty Ltd, http://www.ecotech.com.au/ rainwat.htm.)

Figure 3-2. MicroVol 1100 particulate sampler designed by Ecotech Pty Ltd, Blackburn, Victoria. (Reproduced with permission from Ecotech Pty Ltd, `http://www.ecotech.com.au/uvol1100.htm.`)

Figure 3-3. HV3000 high-volume air sampler designed by Ecotech Pty Ltd, Blackburn, Victoria. (Reproduced with permission from Ecotech Pty Ltd, `http://www.ecotech.com.au/hv3000.htm.`)

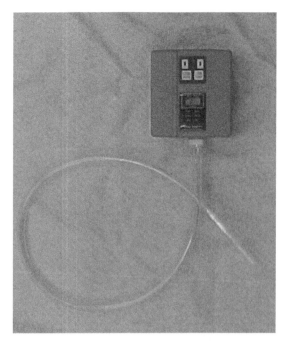

Figure 3-4. Supelco Q-Max pump for taking small samples of organic compounds.

to the laboratory, broken open, extracted into a solvent that effectively desorbs the analytes, and analyzed (usually by gas chromatography or high-performance liquid chromatography). These types of systems are used in industrial workplace settings to monitor the exposure of volatile solvents.

WATER SAMPLING

Water, and the many biota and particles suspended in it, can be somewhat more complicated to sample. First, we look at simple biota samplers. Figure 3-5 shows a plankton net that can be held in place in a stream or towed behind a boat. Water and plankton enter the wide mouth of the net and are funneled toward the narrow collection strainer at the top of the photograph. The mesh size of the netting can be changed to select for different organisms. Figure 3-6 shows a sampling system for macroinvertebrates (mostly, insect larva) attached to bottom materials (rocks, leaves, and sticks). This system is used by selecting the area to be sampled, placing the 1-by-1 foot brace securely over the stream medium, and allowing the water to flow over the sampling area but into the net (the net goes downstream of the sampling area) and brushing the macroinvertebrates off and into the net. After all of the stream medium has been removed, the macroinvertebrates are washed into the end of the net and placed in containers for sorting and identification.

Water (liquid) samplers come in a variety of shapes and sizes suited for a variety of specific purposes. Grab samples of surface waters can be obtained simply by dipping a beaker into water. For hard-to-reach waters or waters/liquids

Figure 3-5. Plankton sampler. (Courtesy of Forestry Suppliers, Inc., `http://www.forestry-suppliers.com.`)

Figure 3-6. Macroinvertebrate sampler for small streams. (Courtesy of Forestry Suppliers, Inc., `http://www.forestry-suppliers.com.`)

that are potentially hazardous, a robotic sample arm can be used (Figure 3-7). Samples can also be taken as a function of depth in a system using automated samplers, such as a van Dorn sampler (Figure 3-8). These samplers work by opening the ends of the unit and restraining them by attaching each end of the tubing to a release mechanism. The unit is lowered to the depth of interest and a messenger (a metal weight) is sent down the connecting rope. The messenger hits the release mechanism and both ends of the unit close, trapping the water inside

Figure 3-7. Robotic arm sampler for grab samples. (Courtesy of Forestry Suppliers, Inc., `http://www.forestry-suppliers.com.`)

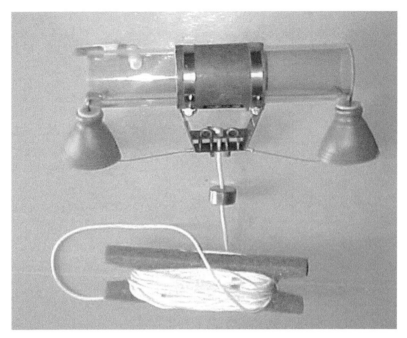

Figure 3-8. Automated water sampler for taking samples as a function of depth.

Figure 3-9. Bailer for taking water samples from a groundwater well. (Courtesy of Forestry Suppliers, Inc., http://www.forestry-suppliers.com.)

the cylinder. These systems can be used individually or as a series of samplers on a single rope.

GROUNDWATER SAMPLING

Groundwater sampling is inherently difficult. The first and most obvious problem is installation of a sampling well in a manner that does not change the integrity of the surrounding water. Once you have convinced yourself that this has been achieved, water can be withdrawn using a simple device such as the water bailer shown in Figure 3-9. This bailer closes each end of the tube when the messenger (the separate metal piece) is dropped along the rope. Some bailers have a ball valve in the bottom that is open as the bailer is lowered into the well and water column. When the bailer is pulled upward, the ball reseals and closes the bottom of the sampler. Thus, water can be taken from specific depths in a groundwater well or tank of water. Pumps are more automated, and expensive, but they may become contaminated during sampling. Bailers are relatively cheap and can be disposed of after each sample is taken, which avoids cross-contamination of wells and storage tanks.

SEDIMENT/SLUDGE SAMPLING

Shallow systems can be sampled using grab samplers such as those shown in Figure 3-10. If a deeper profile is needed, a coring device is used (Figure 3-11).

Figure 3-10. Coring device for shallow water systems. (Courtesy of Forestry Suppliers, Inc., `http://www.forestry-suppliers.com`.)

Figure 3-11. Coring device for shallow water systems. (Courtesy of Forestry Suppliers, Inc., `http://www.forestry-suppliers.com`.)

The coring device contains a metal or plastic tube containing the sample, which can be frozen, sectioned by depth, and extracted for analysis. The sampling of deeper lake systems uses the same type of approach, but the corer is dropped from the boat and retrieved using a rope. Cores as deep as 20 feet have been taken using these devices.

SOIL SAMPLING

Soils are relatively easy to sample and can be collected with samplers as simple as scoops (Figure 3-12). Depth profile samples can be obtained using split-spoon samplers such as those shown in Figures 3-13 to 3-15 or with powered auger systems (Figure 3-16). The sample is easily removed and processed for analysis.

IN-SITU ANALYSIS

Relatively clean water samples can be analyzed in the field using probes and automated water analysis kits. A variety of probes, such as the one shown in Figure 3-17, are available for determination of specific anions, some cations, pH, temperature, salinity, conductivity, dissolved oxygen, selected dissolved gases,

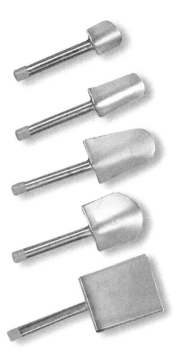

Figure 3-12. Stainless steel scoops used to take surface soil samples. (Courtesy of Forestry Suppliers, Inc., http://www.forestry-suppliers.com.)

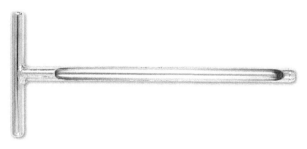

Figure 3-13. Split-spoon sampler for surface samples. (Courtesy of Forestry Suppliers, Inc., http://www.forestry-suppliers.com.)

oxidation–reduction potential, and other parameters. Several portable water analysis kits are available commercially. Two of these are shown in Figures 3-18 and 3-19. Again, these are useful primarily for relatively clean water systems that are not subject to interference. The procedures used by these units are well documented and are very similar to the procedures used in wet/colorimetric chemical analysis.

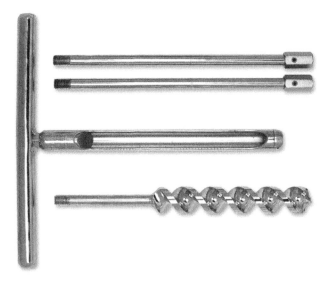

Figure 3-14. Split-spoon sampler used to obtain deeper samples. (Courtesy of Forestry Suppliers, Inc., http://www.forestry-suppliers.com.)

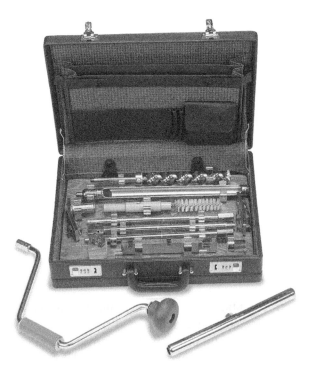

Figure 3-15. Split-spoon sampler with extension rods for deep samples. (Courtesy of Forestry Suppliers, Inc., http://www.forestry-suppliers.com.)

Figure 3-16. Powered auger sampler. (Courtesy of Forestry Suppliers, Inc., `http://www.forestry-suppliers.com.`)

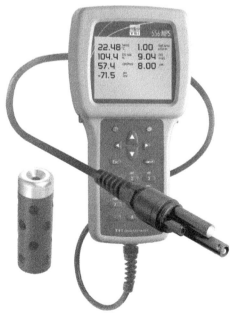

Figure 3-17. Automated probe for in-situ analysis. (Courtesy of Forestry Suppliers, Inc., `http://www.forestry-suppliers.com.`)

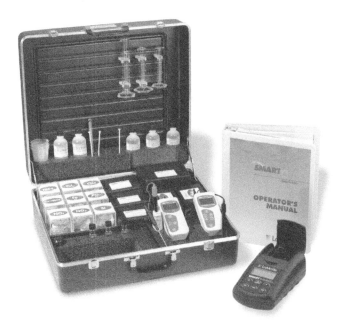

Figure 3-18. Portable water analysis kit. (Courtesy of Forestry Suppliers, Inc., `http://www.forestry-suppliers.com.`)

SAMPLE PRESERVATION AND STORAGE

Finally, after you have taken your sample, you must usually preserve it. The most common way to preserve samples is to cool them to 4°C. Other samples require chemical additions. Your SOPs will clearly outline preservation procedures for your samples. Each state, industry, and federal agency has its own set of sampling, preservation, and storage conditions that must be met if you analyze samples for them.

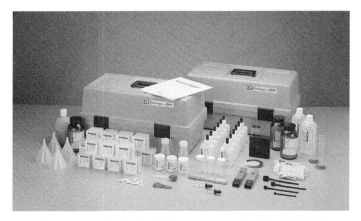

Figure 3-19. Portable water analysis kit. (Photogram provided by Hack Company, `http://www.hach.com.`)

PART 2

EXPERIMENTS FOR AIR SAMPLES

PART 2

EXPERIMENTS FOR USE

4

DETERMINATION OF HENRY'S LAW CONSTANTS

Purpose: To determine Henry's law constants using a gaseous purge technique

To learn the operation of a capillary column gas chromatograph equipped with an electron capture detector

BACKGROUND

The *Henry's law constant* (HLC) is defined as the partial pressure of an analyte divided by its aqueous concentration [equation (4-1)]. This property is important in determining the equilibrium distribution of an analyte between the atmosphere and water solutions, as when raindrops fall through the atmosphere and equilibrate with gases and pollutants. In theory, pure rainwater would have a pH of 7.00, since it is distilled water from the oceans or rivers. But as rain falls through the atmosphere, it equilibrates with CO_2, which is present at a concentration of approximately 380 ppm, depending on where you are on Earth (rural versus industrial locations). The equilibrium concentration of CO_2 in water under atmospheric conditions is described by its Henry's law constant. As noted in Chapter 23, this results in a rainwater pH of 5.5. Similarly, pollutants such as SO_3 (a precursor for acid rain), pesticides, and a variety of hydrocarbons present in the atmosphere can be dissolved in rainwater, and each has a Henry's law constant describing how it partitions.

Another important example is dissolved oxygen, an essential gas for all aerobic aquatic life-forms. The partial pressure in a dry atmospheric sample is about

Environmental Laboratory Exercises for Instrumental Analysis and Environmental Chemistry
By Frank M. Dunnivant
ISBN 0-471-48856-9   Copyright © 2004 John Wiley & Sons, Inc.

0.19 atm. This results in an aqueous equilibrium concentration of approximately 11.3 mg/L at 10°C (and zero salt content), reflecting an HLC for O_2 at 10°C of 0.538 atm·m³/mol. Confirm this by finding the HLC for O_2 in a reference text or on the Internet.

$$\text{HLC} = \frac{\text{partial pressure of analyte (atm)}}{\text{aqueous concentration (mol/m}^3)} \tag{4-1}$$

By knowing any two of the quantities in the HLC equation, you can calculate the remaining variable. This can be very useful since most HLC values are known and either the partial pressure or aqueous solubility will be relatively easy to measure for a given analyte. Also by measuring the partial pressure and aqueous concentration for a given system and by knowing the HLC, you can tell which direction equilibrium is shifting (from the gas or aqueous phase). Approaches such as these have been used by researchers to study the atmosphere–water interactions for triazine (Thurman and Cromwell, 2000; Cooter et al., 2002), miscellaneous pesticides in Greece (Charizopoulos and Papadopoulou-Mourki-dou, 1999), miscellaneous pesticides in the Chesapeake Bay (Harman-Fetcho et al., 2000; Bamford et al., 2002), dioxins and polychlorinated biphenyls (PCBs) in Lake Baikal (Mamontov et al., 2000), and PCBs in the Great Lakes (Subhash et al., 1999). Thus, although the HLC is simple in concept, it is also an important environmental modeling parameter.

THEORY

In 1979, Mackay et al. introduced a new gaseous purge technique for determining Henry's law constants for hydrophobic compounds. This technique has been used to determine HLCs for a variety of pesticides, hydrocarbons, and PCBs. The technique uses a vessel such as the Sherer impinger shown in Figure 4-1. In your experiments, 300 mL of analyte-containing solution will be added to the impinger. One or multiple analytes can be analyzed in a single experiment. The system is sealed with a ground-glass joint, and pure gas is purged through the solution at a rate of 0.500 L/min. The use of ultrapure gas is recommended, but normal-grade gas can be used and purified by placing a Tenax resin tube immediately prior to the gas entry port. As the purge gas enters the system, it passes through a glass frit, and the small bubbles that result equilibrate with the analytes dissolved in the aqueous solution, thus stripping the analytes from the solution. The gas containing the analytes passes through the solution and exits the impinger at the top. A resin tube containing Tenax resin is positioned at the exit port to remove the analytes from the gas stream. The resin tubes are changed with respect to time, thus allowing a time-dependent profile of the removal of analytes from the solution. Subsequently, the Tenax tubes are extracted with acetone, followed by isooctane, which strips the analyte into solution. The isooctane layer is analyzed on a capillary column gas chromatograph equipped with an electron capture detector.

Gas entry
port

Tenax resin
tube location

Purge vessel
containing 300
mL of solution

Purge frit

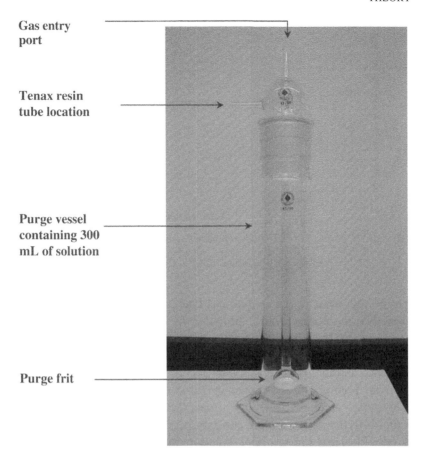

Figure 4-1. Sherer impinger.

There are several basic assumptions that allow calculation of the HLC from the purge experiment. These assumptions include (1) that the volume of water in the impinger does not change significantly during the experiment (2) that equilibrium is established between the aqueous and gas phases before the gas exits the solution (3) that a constant known temperature (isothermal) is used for the purge vessel, and (4) that Henry's law is obeyed over the relevant analyte concentration range. These assumptions can easily be established.

The release of analytes from solution follows a first-order rate law, represented by

$$M_t = M_i - M_i e^{-kt} \qquad (4\text{-}2)$$

where M_t is the mass of analyte purged (ng) at time t, M_i is the total initial mass of analyte, k is the depletion rate (day^{-1} or hour^{-1}), and t is time (days or hour). However, equation (4-2) is used only to monitor the removal of analyte with time

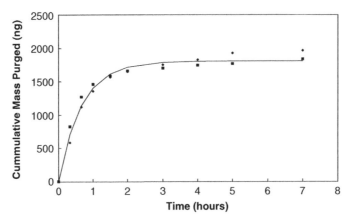

Figure 4-2. Release profile for 2,2′-dichlorobiphenyl from 300.mL of solution in a Sherer impinger.

and to ensure that most of the analyte has been removed (i.e., that a plateau has been reached in the purge profile, allowing the estimation of the total mass of analyte originally in the impinger). Such a purge profile is shown in Figure 4-2 for 2,2′-dichlorobiphenyl. After a stable plateau has been reached, the purge experiment is stopped and the data are analyzed according to Mackay et al. (1979). The raw data from Figure 4-2 are shown in Table 4-1 and are transformed into a $\ln(C/C_0)$ plot [see equation (4-3)] to estimate the depletion rate constant. As seen in equation (4-4), the depletion rate constant is defined as a function of the HLC, gas flow rate, ideal gas law constant, solution volume, and temperature.

$$\ln(C/C_0) = -\text{Dr} \cdot t \tag{4-3}$$

$$\ln(C/C_0) = -(\text{HLC} \cdot G/VRT)t \tag{4-4}$$

where

C = cumulative analyte concentration (mass, ng) removed from the system at time t

C_0 = total analyte concentration (mass, ng) in the original solution (at $t = 0$) (obtained from Figure 4-2)

Dr = depletion rate constant

t = time (days or hours)

HLC = Henry's law constant

G = gas flow rate (0.500 L/min)

V = solution volume (0.300 L)

R = ideal gas law constant (0.08206 L·atm/mol·K)

T = temperature (K)

A linear regression is performed on the time versus $\ln(C/C_0)$ data to obtain the depletion rate constant (slope of the line). In Figure 4-3 this results in a depletion rate constant of 0.879 h^{-1} for 2,2′-DCB. Using equation (4-4) and the experimental conditions given below it, we obtain a HLC for 2,2′-DCB of

TABLE 4-1. Data Used to Generate Figures 4-2 and 4-3

2,2′-DCB Data and Fitted Data

Purge Interval Time (days)	Purge Interval Time (hrs)	Mass in Purge Interval	Mass in Purge Interval	Cummulative Mass in Purge Interval	Cummulative Mass in Purge Interval	C/Co R1	C/Co R2	ln (C/Co) R1	ln(C/Co) R2
0	0	0	0	0	0				
0.01389	0.33336	585	826	585	826	0.6769	0.5442	−0.3903	−0.6085
0.02778	0.66672	535	447	1120	1272	0.3815	0.2975	−0.9636	−1.2124
0.04167	1.00008	237	189	1357	1461	0.2509	0.1931	−1.3826	−1.6447
0.0625	1.5	209	129	1565	1590	0.1357	0.1221	−1.9976	−2.1029
0.08333	1.99992	86.2	75	1652	1665	0.0880	0.0807	−2.4299	−2.5174
0.125	3	104	39.4	1755	1704	0.0308	0.0589	−3.4788	−2.8318
0.16667	4.00008	74	43.6	1829	1748	−0.0100	0.0348		−3.3580
0.20833	4.99992	99.9	23.5	1929	1771	−0.0652	0.0218		−3.8243
0.29167	7.00008	39.5	67.5	1968	1839	−0.0870	−0.0154		

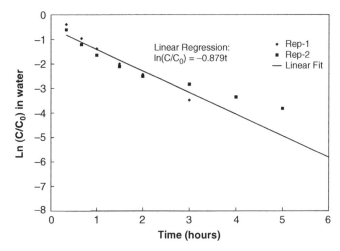

Figure 4-3. Linear transformation of data to obtain the depletion rate constant (Dr).

2.15×10^{-4} atm·m^3/mol, which is in good agreement with literature values $(2.19 \times 10^{-4}$ to $5.48 \times 10^{-4})$.

ACKNOWLEDGMENT

I would like to thank Josh Wnuk (Whitman College, Class of 2003) for data collection and analysis.

REFERENCES

Bamford, H. A., F. C. Ko, and J. E. Baker, *Environ. Sci. Technol.*, **36**(20), 4245–4252 (2002).

Charizopoulos, E. and E. Papadopoulou-Mourkidou, *Environ. Sci. Technol.*, **33**(14), 2363–2368 (1999).

Cooter, E. J., W. T. Hutzell, W. T. Foreman, and M. S. Majewski, *Environ. Sci. Technol.*, **36**(21), 4593–4599 (2002).

Harmon-Fetcho, J. A., L. L. McConnell, C. R. Rice, and J. E. Baker, *Environ. Sci. Technol.*, **34**(8), 1462–1468 (2000).

Mackay, D., W. Y. Shiu, and R. P. Sutherland, *Envion. Sci. Technol.*, **13**(3), 333–337 (1979).

Mamontov, A. A., Mamontova, E. A., and E. N. Tarasova, *Environ. Sci. Technol.*, **34**(5), 741–747 (2000).

Subhash, S., R. E. Honrath, and J. D. W. Kahl, *Environ. Sci. Technol.*, **33**(9), 1509–1515 (1999).

Thurman, E. M. and A. E. Cromwell, *Environ. Sci. Technol.*, **34**(15), 3079–3085 (2000).

IN THE LABORATORY

During the first laboratory period, you will prepare your purge apparatus (Sherer impinger) and during the following 24 hours take samples to determine the Henry's law constant for selected pesticides and PCBs. Your samples (Tenax resin tubes) can be extracted as you take them or during the beginning of the next laboratory period. In the second laboratory period you will analyze the sample extracts on the gas chromatograph and process your data.

Safety Precautions

- Safety glasses must be worn at all times during this laboratory experiment.
- Most if not all of the compounds you will use are carcinogens. Your instructor will prepare the aqueous solution of these compounds so that you will not be handling high concentrations. The purge solution you will be given contains parts per billion (ppb)-level concentrations and is relatively safe to work with. You should still use caution when using these solutions since the pesticides and PCBs are very volatile when placed in water. Avoid breathing the vapors from this solution.
- Extracts of the Tenax tubes should be conducted in the hood since you will be using acetone and isooctane, two highly flammable liquids.

Chemicals and Solutions

Neat solutions of the following compounds will be used by your instructor to prepare your aqueous solution:

- 2,2'-Dichlorobiphenyl
- Lindane
- 4,4'-Dichlorobiphenyl
- 2,2',6,6'-Tetrachlorobiphenyl
- Aldrin
- 2,2',4,4',6,6'-Hexachlorobiphenyl
- 3,3',4,4'-Tetrachlorobiphenyl
- Dieldrin
- 4,4'-DDD (dichlorodiphenyldichloroethane)
- 4,4'-DDT (dichlorodiphenyltrichloroethane)
- Methoxychlor
- Endosulfan I (not added to purge system, but used as a GC internal standard)

You will need, in addition:

- Tenax resin, chromatography grade
- Deionized water

Equipment and Glassware

- Sherer impingers (one per student group) (available from Ace Glassware; use the frit that allows gas to exit at the bottom of the impinger)
- Pasteur pipets filled with Tenax resin
- ~15-mL glass vials equipped with a Teflon-lined septum (12 per Sherer impinger setup or student group)
- Tygon tubing
- Brass or stainless steel fine metering valves
- Brass or stainless steel tees

PROCEDURE

In the lab, the Sherer impinger will already be set up and the purge solutions prepared. Your instructor will go over the setup and show its proper operation (Figure 4-4). Before you start the experiment, you will need to prepare Tenax resin sampling tubes. Tenax is a resin that has a high affinity for hydrophobic compounds and will absorb them when water or gas containing analytes is passed through the resin. Prepare the tubes by taking a glass Pasteur pipet and filling the narrow end with a small amount of glass wool. Next, place the Tenax resin tube in the pipet, leaving enough room for more glass wool at the constriction. This will leave about 1 to 2 cm of empty space at the top of the pipet (we will need this to add solvent to the pipet to desorb the analytes later). Clean the Tenax resin traps by passing at least 5 mL of pesticide-grade acetone through it, followed by 5 mL of pesticide-grade isooctane. Dry the tubes by placing them in the gas stream of the Sherer impinger (with no analyte present). You will need 14 tubes per Sherer impinger unless you desorb the tubes as you collect them.

If this is the case, you need only two tubes but you must still dry the tubes between samples. Tenax resin tubes should be wrapped and stored in aluminum foil.

1. Set up the impinger as shown by your instructor and set the gas flow rate while the flask is filled with deionized water (no analyte solution) (this will be a good time to purge the solvent from the Tenax purge tubes). Leave the final tube on the setup.

2. Leave the gas flow set as adjusted in step 1, but disassemble the apparatus and empty the flask.

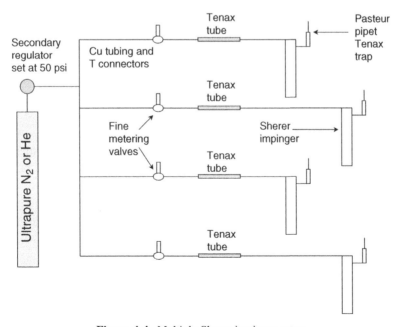

Figure 4-4. Multiple Sherer impinger setup.

3. Fill the flask with 300 mL of analyte-containing water.

4. Have a stopwatch or clock ready, assemble the Sherer flask, turn the ground-glass joint tightly to ensure a seal, and note the time. This is $t = 0$.

5. Check the flow rate and if needed, adjust it to 0.500 L/min.

6. Sample at the following times to obtain a complete purge profile:

20 minutes

40 minutes

1.00 hour

1.50 hours

2.00 hours

3.00 hours

4.00 hours

5.00 hours

7.00 hours

17.0 hours

29.0 hours

Desorbing the Tenax Resin Tubes

7. Place the Tenax resin tube in a small clamp attached to a ring stand. Lower the tube so that it just fits into a ~15-mL glass vial.

8. Pipet 5.00 mL of pesticide-grade acetone onto the top of the Tenax resin trap. Allow the acetone to reach the top of the resin with gravity. You may have to apply pressure with a pipet bulb to break the pressure lock caused by bubbles in the tube, but be careful not to blow more air into the tube. After the second or third application (with a bulb) the acetone should flow with gravity. (The reason for adding acetone is to remove any water from the resin tube that will not mix or be removed by the hydrophobic isooctane.)

9. Pipet 5.00 mL of pesticide-grade isooctane onto the resin trap. After the isooctane has passed through the resin trap, force the remainder of the isooctane out of the pipet with a bulb. Remove the vial from below the tube, being careful not to spill any of the contents.

10. Add 10.0 mL of deionized water to the extraction vial and 0.25 g of NaCl. (NaCl will break any emulsion that forms in the solvent extraction step.)

11. Add 8.0 μL of a 32.70-ppm Endosulfan I (in isooctane) that your instructor will have prepared for you. Endosulfan I will act as an internal standard for the gas chromatographic (GC) analysis.

12. Seal the vial and shake it vigorously for 30 seconds. Allow the layers to separate, transfer 1 to 2 mL of the top (isooctane) layer into a autoinjection vial, and seal it.

13. Add your name to the GC logbook and analyze the samples using the following GC conditions:

    1.0-μL injection

    Inlet temperature = 270°C

    Column:
    HP-1 (cross-linked methyl silicone gum)
    30.0 m (length) by 530 μm (diameter) by 2.65 μm (film thickness)
    4.02-psi column backpressure
    3.0-mL/min He flow
    31-cm/s average linear velocity

    Oven:
    Hold at 180°C for 1.0 minute
    Ramp at 5.0°C/min
    Hold at 265°C for 16.0 minutes
    Total time = 34.0 minutes

    Detector:
    Electron-capture detector
    Temperature = 275°C
    Makeup gas = Ar with 1 to 5% CH_4
    Total flow = 60 mL /min

    A sample chromatogram is shown in Figure 4-5. Calibration standards will be supplied by your instructor and will range in concentrations from 1.00 to 500 ppb. Approximate retention times for the given GC setting are as follows:

Analyte	Elution Time (min)	Analyte	Elution Time (min)
2,2′-DCB	9.63	Endosulfan I (IS)	19.75
Lindane	12.13	Dieldrin	20.95
4,4′-DCB	12.71	DDD	22.20
2,2′6,6′-TCB	13.82	DDT	24.72
Aldrin	16.86	Methoxychlor	28.33
2,2′,4,4′,6,6′-TCB	18.86		

14. Sign out of the GC logbook and note any problems you had with the instrument.

15. Analyze the data and calculate the HLC for all the compounds in your samples.

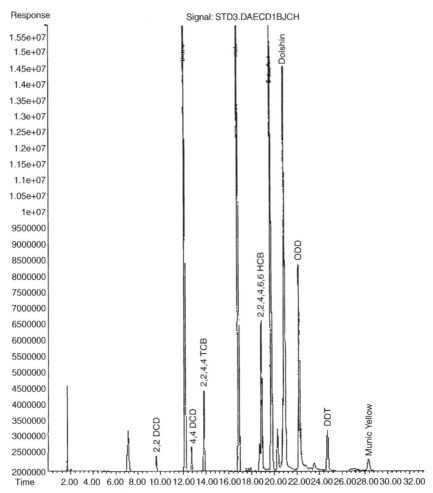

Figure 4-5. Output from the GC.

Waste Disposal

The water remaining in your Sherer impinger has been purged of all analytes and can be disposed of down the drain. Your sample extracts must be treated as hazardous waste since they contain acetone, isooctane, and chlorinated hydrocarbons. These should be placed in a glass storage container and disposed of in accordance with federal guidelines.

ASSIGNMENT

1. Turn in a diagram of your purge setup.
2. Turn in a spreadsheet showing the HLC calculation.
3. Compare the HLC values calculated to values from the literature.

ADVANCED STUDY ASSIGNMENT

1. Draw and describe each major component of a basic capillary column gas chromatograph.
2. Calculate the Henry's law constant with the data set in Table 4-2 for Dieldrin:

$$\text{Purge gas flow rate} = 0.500 \, \text{L/min}$$
$$\text{System temperature} = 25°C$$
$$\text{Total mass of Dieldrin in flask } (C_0) = 725 \, \text{ng}$$
$$\text{Volume in Sherer impinger} = 300 \, \text{mL}$$

   Mass in each purge interval is in measured in nanograms. Express your answer in $\text{atm} \cdot \text{m}^3/\text{mol}$.
3. Compare your answer to the value from a reference text or a value from the Internet.

TABLE 4-2. Sample Data Set

Purge Interval Time (days)	Purge Interval Time (hrs)	Mass in Purge Interval R-1	Mass in Purge Interval R-2	Cummulative Mass in Purge Interval R-1	Cummulative Mass in Purge Interval R-2	C/Co R1	C/Co R2	ln (C/Co) R1	ln(C/Co) R2
0.01389		65.16	66.5						
0.02778		77.76	74.8						
0.04167		73	71.9						
0.0625		72.8	75.2						
0.08333		69.9	70						
0.125		86.9	84.5						
0.16667		80.7	69.8						
0.20833		76.1	61.6						
0.2917		77.5	63						

DATA COLLECTION SHEET

DATA COLLECTION SHEET

DATA COLLECTION SHEET

DATA COLLECTION SHEET

DATA COLLECTION SHEET

5

GLOBAL WARMING: DETERMINING IF A GAS IS INFRARED ACTIVE

Purpose: To learn to use an infrared spectrophotometer

To determine if a gas is infrared active

BACKGROUND

Although global warming has drawn growing political attention in recent decades, relatively few people understand its causes and implications. Global warming has two faces, one that benefits us and another that may cause serious environmental and economic damage to the planet. Conditions on Earth would be very different without the *greenhouse effect* of atmospheric warming. Natural atmospheric gases, including carbon dioxide and water vapor, are responsible for adjusting and warming our planet's atmosphere to more livable conditions. In fact, there is one popular theory that the Earth is actually a living organism and that under normal conditions (without human interference), the Earth will maintain the life-sustaining environment that it has acquired over the last 100 million years or so. This theory is the *Gaia hypothesis* proposed by James Lovelock, and there are several short books on the subject.

The bad side, the anthropogenic side, of global warming is still strongly debated between some politicians and scientists, but it is generally well accepted among scientists that humans are contributing exponentially to the warming of the planet. Unfortunately, some governments and political parties side with the

Environmental Laboratory Exercises for Instrumental Analysis and Environmental Chemistry
By Frank M. Dunnivant
ISBN 0-471-48856-9   Copyright © 2004 John Wiley & Sons, Inc.

economists, who often have little knowledge of the science behind the argument but are concerned primarily with constant economic growth rather than sustained growth. This bad side to global warming has been studied for several decades and data from these studies is presented below.

First, it is important to understand the nature of the light coming from our Sun to the Earth. Figure 5-1 shows three representations of the wavelengths and intensity of light coming from the surface of the Sun (at 5900 K). The upper dashed line represents the wavelengths and intensity of light as predicted by physicists for a blackbody residing at the temperature of the sun. This line predicts fairly accurately the spectrum of wavelengths observed just outside the Earth's atmosphere by satellites (represented by the upper solid line). The remaining line (the lower solid line) shows the spectrum of wavelengths detected at the Earth's sea surface using similar satellites. As you can see, some of the intensity is reduced and a few of the wavelengths are removed completely by atmospheric gases. The wavelengths in Figure 5-1 are given in micrometers, with ultraviolet (UV) radiation between 0 and 0.3 on the x axis, visible light from 0.3 to about 0.8 and near-infrared (IR) from about 0.8 to the far right side of the plot. As you see, most of the solar radiation entering Earth's atmosphere is in the form of visible light and near-IR radiation.

Next, notice the difference between the UV radiation intensity outside the atmosphere and at sea level. These wavelengths, which cause damage to skin and other materials, are removed in the stratosphere during the formation of ozone shown below (diatomic oxygen absorbs these wavelengths, splits into free oxygen

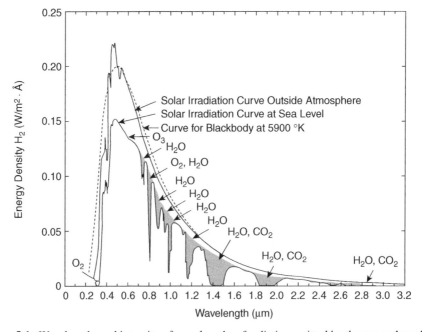

Figure 5-1. Wavelengths and intensity of wavelengths of radiation emitted by the sun and reaching Earth's sea surface. (From Department of the Air Force, 1964.)

radicals, and binds to another O_2 to form O_3). This is the source of concern with chlorofluorohydrocarbons, which interfere with this process and promote the destruction of O_3, thus allowing more high energy UV to reach Earth's surface.

$$O_2(g) + h\nu \rightarrow 2O_2(g)$$
$$O_2(g) + O_2(g) + M \rightarrow O_3(g) + M^*(g) + \text{heat}$$

Visible light is also attenuated significantly by Earth's atmosphere, but not to the extent that it limits the growth of plant life. Some of the visible light is simply absorbed and rereleased as heat in the atmosphere. Other visible wavelengths are scattered and reflected back into space, which is why the astronauts can see the Earth from space. Several compounds in the atmosphere partially or completely absorb wavelengths in the near-IR radiation on the left side of the figure. Absorption of these wavelengths is represented by the shaded areas for O_3, H_2O, O_2, and CO_2. This is one mechanism of global warming, in which the atmosphere is heated by IR radiation incoming from the Sun rather than reradiated from Earth's surface. To fully understand the importance of these gases in global warming, we must also look at the type of radiation the Earth is emitting.

As visible light reaches Earth's surface, it is absorbed by the surface and transformed into heat. This heat is reemitted back into the atmosphere and space by Earth. When physicists estimate the wavelengths and intensity of wavelengths for Earth as a blackbody at 320 K, the dashed-line spectrum shown in Figure 5-2 results. Note that the wavelengths released by Earth are much longer wavelength (far, far to the right of the wavelengths shown in Figure 5-1). These far-infrared

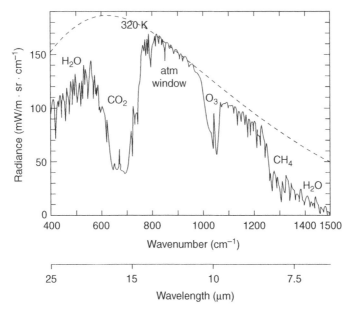

Figure 5-2. Wavelengths and intensity of wavelengths of radiation emitted by the Earth. (From Hanel et al., 1972.)

wavelengths are very susceptible to being absorbed by atmospheric gases, as indicated by the decrease in intensity shown by the solid line. The solid line shows the wavelength and intensity of wavelengths measured by a satellite above Earth's surface, but this time the satellite is pointed at Earth instead of the Sun. Note the strong absorbance by atmospheric constituents, primarily water, methane, and carbon dioxide. By absorbing the IR radiation instead of letting it pass freely into space, the gases heat Earth's atmosphere. The amount of global warming resulting from the reflected IR radiation is related directly to the concentration of atmospheric gases that can absorb the emitted IR radiation. Before we can evaluate the cause of the "bad" global warming, we must look at historical data on concentrations of greenhouse gases (IR-active gases) in the atmosphere.

In the 1950s the U.S. government initiated a project to collect baseline data on planet Earth. One of the most important studies was to monitor the concentration of CO_2 in a remote, "clean" environment. The site selected for this monitoring program was the observatory on Mauna Loa in Hawaii. This site was selected for its location in the middle of the Pacific, away from major pollution sources, and for its high altitude (about 14,000 feet). Data from this monitoring program are shown in Figure 5-3 and are available from the LDEO Climate Data Catalog, which is maintained by the International Research Institute at Columbia University (http://www.ingrid.ldgo.columbia.edu/). Data from 1958 to the year 2000 (not shown) consistently show an increase in atmospheric CO_2 concentrations. In addition, for the first time we can actually see the Earth "breath," as indicated in the inset in Figure 5-3: In the summer, when plant growth is highest in the northern hemisphere, CO_2 levels are at a minimum. This is followed by fall, when plant growth is subsiding and dying, and CO_2 levels start to increase. The CO_2 concentration reaches a maximum in winter, followed by a decrease in spring as plants start growing again to repeat the cycle.

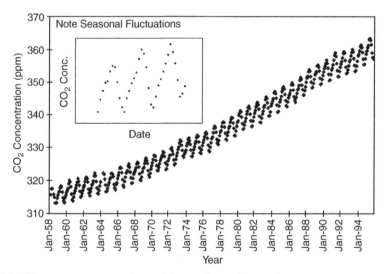

Figure 5-3. CO_2 measurements from Mauna Loa. (Data from http://ingrid.ldgo. columbia.edu/.)

One problem with the data set from Mauna Loa is that it represents only a small snapshot in time; with issues such as global warming, we must look at long-term geological time scales. To do this, scientists have collected ice cores from a variety of places across the Earth. Ice cores represent a long history of atmospheric data. As snow falls over cold areas and accumulates as snow packs and glaciers, it encapsulates tiny amounts of atmospheric gases with it. When ice cores are taken and analyzed carefully, they can give information on the composition of the atmosphere at the time the snow fell on the Earth. An example of these data for the Vostok ice core is shown in Figure 5-4. This data set goes back in time 160,000 years (from left to right) from the present and gives us a long-term idea of the composition of the atmosphere. The three figures show the concentration of CH_4 with time (Fig. 5.4a), the concentration of CO_2 with time (Fig. 5.4b), and the estimated temperature with time (Fig. 5.4c). The CH_4 and CO_2 data are self-explanatory and are simply the gases trapped in the glacier, but the temperature data are a bit more complicated. To estimate the temperature as a function of time, scientists look at the abundance of the oxygen-18 isotope in glacial water. Water on Earth contains mostly oxygen-16, but a small amount of oxygen-18 is present. During warmer geologic times on Earth, more water containing ^{18}O is evaporated from the oceans and falls as snow over cold regions. In contrast, cooler geologic times will have less ^{18}O in the atmospheric and snow. By conducting experiments we can estimate how much ^{18}O is present at a given temperature and estimate what the temperature was when each layer of the glacial water was deposited. This allows Figure 5-4c to be created. When the three figures are compared, a strong correlation between high CH_4 concentrations, high CO_2 concentrations, and high temperature is noticed. This can be understood by

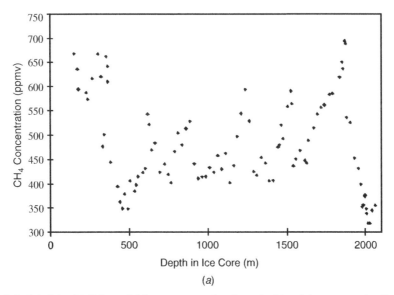

Figure 5-4. (a) CH_4, (b) CO_2, and (c) temperature data from the Vostok ice core study. (Data from `http://ingrid.ldgo.columbia.edu/`.)

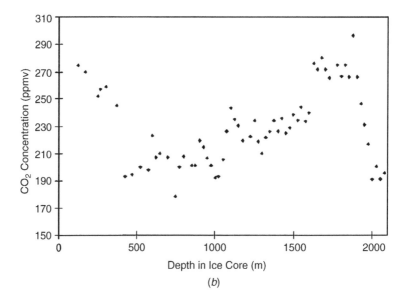

(*b*)

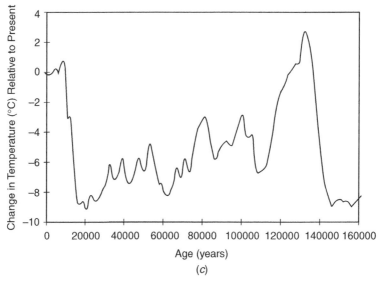

(*c*)

Figure 5-4. (*Continued*)

returning to Figures 5-1 and 5-2 and noting which gases absorb or trap energy in Earth's atmosphere.

Now we combine the CO_2 data from the Vostok ice core and the Mauna Loa data set to create Figure 5-5. Note in the figure that the direction of time changes, going back in time from left to right. This figure contains data going back 160,000 years, and we notice two distinct spikes in CO_2 concentration (and in temperature if we look again at Figure 5-4). The important point to note in Figure 5-5 is the rate at which the CO_2 (and temperature) has changed over time. The natural

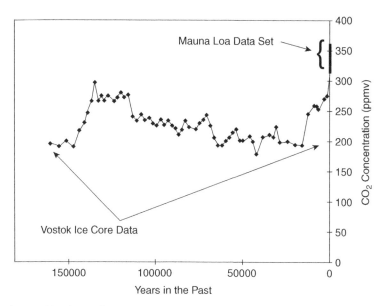

Figure 5-5. Combined data from the Vostok ice core and the Mauna Loa studies. Note the rapid change in CO_2 levels during the present time. The Mauna Loa data are from Keeling (1995, 1996); the Vostok ice core data are from Barnola et al. (1987), Genthon et al. (1987), and Jouzel et al. (1987). (Data from `http://ingrid.ldgo.columbia.edu/`.)

change in CO_2 around 130,000 years ago took more than 30,000 years to go from the lowest to the highest concentration. Similarly, the recent climb in temperature took approximately 20,000 years to reach its current level. This is in contrast to the drastic rate of change that is present in the Mauna Loa data set. This 50-ppm change in CO_2 concentration has occurred in only 50 years, and most predictions of future atmospheric CO_2 concentrations (if we continue to consume petroleum products at current rates) are in the range 700 to 800 ppm by the year 2100 (locate this point in Figure 5-5). This is the global warming that concerns us directly. Some people call for more study of the problem and wish to maintain our use of fossil fuels to preserve our economic status, but based on the data presented here, this is one experiment that we may not wish to conduct.

Although many scientists accept that global temperatures are rising, they are less in agreement about the effects of global warming. Most, however, agree on the following predictions:

- Warmer temperatures (averaging 5 to 10°C by the year 2100)
- Loss of coastal areas to flooding
- Damage to coral reefs (bleaching)
- Increased incidence of violent weather
- Increased outbreaks of diseases (new and old)
- Changing regional climates (wetter or drier, depending on where you live)

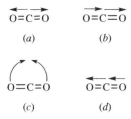

Figure 5-6. Vibrational structures for CO_2.

THEORY

In the background section we saw which greenhouse gases absorb IR radiation and at what wavelengths. But what actually makes a gas IR active? There are two prerequisites for a gas to be IR active. First, the gas must have a permanent or temporary dipole. Second, the vibration of the portion of the molecule having the dipole must be at the same frequency as the IR radiation that is absorbed. When these two criteria are met, the gas molecule will absorb the radiation, increase its molecular vibrations, and thus retain the heat in the atmosphere. This is why gases such as O_2 and N_2 are not IR active; they do not have permanent or sufficiently temporary dipoles. Molecules such as chlorofluorocarbons (CFCs), on the other hand, have permanent dipoles and are very IR active (actually, this is the *only* connection between global warming and ozone depletion—CFCs are active in both cases). However, what about symmetrical molecules such as CO_2 and CH_4? To understand how these molecules are IR active, we must draw their molecular structures.

Figure 5-6 shows several possible vibrational structures for CO_2. The arrows indicate the direction of the stretch. Figure 5-6*a* is the normal way we think about CO_2, with each carbon–oxygen bond stretching in unison and away from the central carbon atom and no dipole present in the molecule. However, the stretches in Figure 5-6*b*, *c*, and *d* are also possible and result in a temporary dipole that can absorb IR radiation. Similar observations can be made for methane (Figure 5-7). The symmetrical orientation is shown in Figure 5-7*a*, while asymmetrical molecules are shown in Figure 5-6*b* and *c*, which contain temporary dipoles. The latter two molecules absorb IR radiation and result in a heating of the atmosphere.

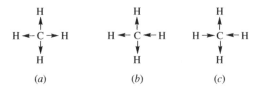

Figure 5-7. Molecular vibrations for methane.

ACKNOWLEDGMENT

I would like to thank Dr. Paul Buckley for taking the IR readings given in the instructor's version of this manual.

REFERENCES

Barnola, J. M., D. Raynaud, Y. S. Korotkevich, and C. Lorius, *Nature*, **329**, 408–414 (1987).

Berner, E. K. and R. A. Berner, *Global Environment: Water, Air, and Geochemical Cycles*, Prentice Hall, Upper Saddle River, NJ, 1996, p. 32.

Department of the Air Force, *Handbook of Geophysics and Space Environmental*, 1965, p. 16–2.

Genthon, C., J. M. Barnola, D. Raynaud, C. Lorius, J. Jouzel, N. I. Barkov, Y. S. Korotkevich, and V. M. Kotlyakov, *Nature*, **329**, 414–418 (1987).

Hanel, R. A., B. J. Conrath, V. G. Kunde, C. Prabhakara, I. Revah, V. V. Salomonson, and G. J. Wolfrod, *J. Geophys. Res.*, **77**(15), 2629–2641 (1972).

Houghton, J. T., F. J. Jenkins, and J. J. Ephraums (eds.), *Climate Change: The IPCC Scientific Assessment*, Cambridge University Press, Cambridge, 1990.

Houghton, J. T., L. G. Meira Filho, B. A. Callander, N. Harris, A. Katterberg, and K. Maskell (eds.) *Climate Change: The Science of Climate Change, The IPCC Scientific Assessment*, Cambridge University Press, Cambridge, 1995.

Jager, J. and F. L. Ferguson (eds.), *Climate Change: Science, Impacts, and Policy*, Proceedings of the 2nd World Climate Conference, Cambridge University Press, Cambridge, 1991.

Jouzel, J., C. Lorius, J. R. Petit, C. Genthon, N. I. Barkov, V. M. Kotlyakov, and V. M. Petrov, *Nature*, **329**, 403–408 (1987).

Keeling, C. D., T. P. Whorf, M. Wahlen, and J. van der Plicht, *Nature*, **375**, 666–670 (1995).

Keeling, C. D., J. F. S. Chine, and T. P. Whorf, *Nature*, **382**, 146–149 (1996).

LDEO Climate Data Catalog, maintained by International Research Institute (IRI) at Columbia University, `http://www.ingrid.ldgo.columbia.edu/`.

Mintzer, I. M. (ed.), Stockholm Environmental Institute, *Confronting Climate Change: Risks, Implications, and Responses*, Cambridge University Press, Cambridge, 1992.

Skoog, D. A., F. J. Holler, and T. A. Nieman (eds.), *Principles of Instrumental Analysis*, 5th ed., Saunder College Publishing, Philadelphia, 1998.

World Resources Institute, *World Resources, 1996–1997*, Oxford University Press, Oxford, 1996.

IN THE LABORATORY

There is no exact procedure for conducting this laboratory other than consulting the users' guide for your IR instrument. Sign in the instrument logbook and remember to record any problems with the instrument when you finish. You will be provided with a variety of gases that you will measure on your IR instrument. Print out the spectrum for each gas and use the resources in your library to determine what type of vibration is occurring at each wave number where you observe absorption of IR radiation.

Safety Precautions

- Avoid the use of methane or other flammable gases around electronic equipment or flames.

Chemicals

- *Gases*: N_2, O_2, a CFC, a CFC substitute, CO_2, and CH_4

Equipment

- IR spectrophotometer
- IR gas cell

Waste Disposal

The gas cells should be filled and emptied in a fume hood.

ASSIGNMENT

Turn in your IR spectrum and label each peak with respect to the vibration that is occurring.

ADVANCED STUDY ASSIGNMENT

1. What are the requirements for a gas to be IR active?
2. Look up the composition of Earth's atmosphere. Which gases would you expect to be IR active?
3. Draw a diagram of a basic IR instrument and explain how it works.
4. Using the Internet, find how much CO_2 is emitted each year by the most productive nations. Which nation has the largest emissions?

DATA COLLECTION SHEET

DATA COLLECTION SHEET

DATA COLLECTION SHEET

DATA COLLECTION SHEET

6

MONITORING THE PRESENCE OF HYDROCARBONS IN AIR AROUND GASOLINE STATIONS

Purpose: To determine the exposure of citizens to gasoline vapors

To learn to use a personal sampling pump

To learn to analyze gasoline components on a gas chromatograph

BACKGROUND

Each day we are exposed to a variety of organic vapors. Yet we experience perhaps the greatest level of exposure when we fill our automobiles with gasoline. Gasoline contains a variety of alkanes, alkenes, and aromatics. In California alone, it has been estimated that 6,100,000 lb of gasoline vapors per year are released into the atmosphere (http://www.arb.ca.gov). It is also interesting to note that at least 23 of the 1430 National Priorities List sites (compiled by the U.S. Environmental Protection Agency) contain automotive gasoline (http://www.atsdr.cdc.gov).

Table 6-1 shows the approximate composition of unleaded gasoline. You should note several carcinogens in this list. The right-hand column shows data on exposure limits (http://www.bpdirect.com); the allowed concentrations shown are relatively high compared to some pollutant exposures, but if you consider how often you (or the gas station attendant) are exposed to these vapors, you may start to appreciate the problem and potential cancer risk.

Environmental Laboratory Exercises for Instrumental Analysis and Environmental Chemistry
By Frank M. Dunnivant
ISBN 0-471-48856-9   Copyright © 2004 John Wiley & Sons, Inc.

TABLE 6-1. Composition of Unleaded Gasoline

Component	Percent Range by Weight	Exposure Limits (ppm)
Benzene	0–3	1–5
Butane	4–6	800
Cyclohexane	0–1	300
Ethylbenzene	0–2	100–125
Heptane	6–8	400–500
Hexane	6–8	50–500
Pentane	9–11	600–1000
Toluene	10–12	100–200
Trimethylbenzene	0–3	25
Xylene	8–10	100

Source: http://www.bpdirect.com

But what exactly are the risks of exposure? Laboratory animals (rats and mice) exposed to high concentrations of gasoline vapors (at 67,262 and 2056 ppm) showed kidney damage and cancer of the liver. n-Heptane and cyclohexane can cause narcosis and irritation of the eyes and mucous membranes. In studies using rabbits, cyclohexane caused liver and kidney changes. Benzene, a known human carcinogen, has an eight-hour exposure limit of 0.5 ppm. Studies have shown that exposure to benzene vapor induce leukemia at concentrations as low as 1 ppm. Trimethylbenzene (isooctane) has an eight-hour exposure limit of 25 ppm and above this limit can cause nervousness, tension, and anxiety as well as asthmatic bronchitis. n-Hexane has been shown to cause peripheral nerve damage and hexanes show narcotic effects at 1000 ppm. Toluene can cause impairment of coordination and momentary memory loss at 200 to 500 ppm. Palpations, extreme weakness, and pronounced loss of coordination can occur at 500 to 1500 ppm. The eight-hour exposure limit for toluene is 100 ppm. (Data in this paragraph were obtained from http://www.brownoil.com.)

As you can see from the discussion above, exposure to gasoline vapors, although routine, should be of concern to anyone filling his or her automobile's gas tank.

THEORY

The sampling of gasoline vapors is a relatively easy process. Figure 6-1 shows a typical sampling pump and sample cartridge. The pump comes calibrated from the factory with respect to airflow, and the flow can be adjusted on most pumps. The pump pulls the air and vapors through the sampling tube, thus avoiding both contamination of the sample tube with compounds from the pump and contamination of the sampling pump with gasoline vapors. A variety of sample tubes are available, with difference resins designed for efficient adsorbance of analytes of

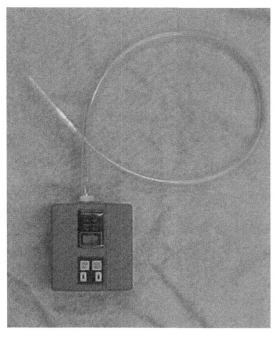

Figure 6-1. Q-Max personal sampling pump. (Supelco, Inc.)

interest. The tube you will use is filled with fine-grained charcoal. Each tube contains two compartments of resin. The large compartment is at the end where the vapors are drawn into the system. The air then passes through a smaller compartment, which is analyzed separately to see whether vapors have saturated the first compartment of resin and passed to the second compartment. When this saturation occurs, it is referred to as breakthrough, and the sample is not usable, since you do not know if vapor has also passed the second tube. The only difficult task in designing a sampling procedure is to determine how long to sample to trap enough vapors to analyze on the gas chromatograph. Your instructor will specify how long you should sample but typically a 5 to 10 minute sample will suffice. You will also be using decane as an internal standard for the GC. Your instructor will review the use of this approach at the beginning of the laboratory.

REFERENCES

http://www.arb.ca.gov, accessed Oct. 5, 2003.

http://www.atsdr.cdc.gov, accessed Oct. 5, 2003.

http://www.bpdirect.com, accessed Oct. 5, 2003.

http://www.brownoil.com, accessed Oct. 5, 2003.

http://www.cdc.gov/niosh/homepage.html, accessed Oct. 5, 2003.

IN THE LABORATORY

You will be divided into groups and sent to a local gasoline station to take samples. Your instructor will have already contacted the owner of the station and asked for permission. You may actually fill cars with gasoline, or you may simply stand beside car owners (or station attendants) as they operate the pumps. Next, you will extract the samples and analyze them on the GC. There are many compounds present in gasoline, but we will only be analyzing selected compounds.

Safety Precautions

- Safety glasses must be worn when in the laboratory.
- All of these vapors have exposure limits, and many are carcinogens. Avoid exposure to these vapors in the laboratory by working in fume hoods. Your instructor may choose to use carbon disulfide, a highly toxic and cancer-causing agent. *Always* work in the fume hood with this solvent, even when filling the syringe for injection into the GC.

Chemicals and Solutions

We will analyze for the compounds shown in Table 6-2. Decane will be used as the internal standard that will be added to your desorption (extraction) solvent (pentane or carbon disulfide) as well as the GC calibration standards at a concentration of 29.2 ppm. You will use the density to calculate the concentration in your calibration standards (volume added times density equals mass added to volumetric).

Use the data shown in Table 6-3 to prepare your GC calibration standards if these standards are not provided from the stockroom. The solvent used for your samples and standards will be pentane or carbon disulfide containing the same concentration of decane as used in the calibration standards. You will also need approximately 50 mL of internal standard solution for extraction of your samples from the charcoal. Your instructor may also have this solution prepared.

TABLE 6-2. Density of Compounds to Be Used in Calibration Standards

Compound	Density (g/mL or mg/mL)	Compound	Density (g/mL or mg/mL)
Benzene	0.8787	*m*-Xylene	0.8684
Ethyl benzene	0.866	*o*-Xylene	0.8801
n-Heptane	0.684		
Isooctane	0.6919	Decane	0.73 (internal standard)
Toluene	0.866		

TABLE 6-3. Solutions for Making Calibration Standards from Pure (Neat) Compounds[a]

Compound	μL Neat to a 25-mL Vol.	Mass (mg) in Vol.	Stock Conc. in 25-mL Vol.	Std. 1 1:50 Dilution, Then 1:10	Std. 2 1:50 Dilution Then 2:10	Std. 3 1:50 Dilution, Then 5:10	Std. 4 1:50 Dilution	Std. 5 1:25 Dilution
Benzene	50	43.935	1757.4 ppm	3.5148	7.0296	17.574	35.148	70.296
Ethyl Benzene	50	43.3	1732	3.464	6.928	17.32	34.64	69.28
n-Heptane	50	34.2	1368	2.736	5.472	13.68	27.36	54.72
Isooctane	50	34.595	1383.8	2.7676	5.5352	13.838	27.676	55.352
Toluene	50	43.3	1732	3.464	6.928	17.32	34.64	69.28
m-Xylene	50	43.42	1736.8	3.4736	6.9472	17.368	34.736	69.472
o-Xylene	50	44.005	1760.2	3.5204	7.0408	17.602	35.204	70.408
Decane	50	36.5	1460	29.2	29.2	29.2	29.2	29.2

[a] Use the densities shown in Table 6-2. Concentrations of analytes in standards 1–5 are in ppm. Note that the concentrations of Decane should be the same (29.2 ppm) in all standards and sample extracts.

TABLE 6-4. Approximate Retention Times for Analytes on a DB-1 Column

Analyte	Retention Time (min)	Analyte	Retention Time (min)
Benzene	4.52	Toluene	8.05
Ethyl Benzene	10.67	m-Xylene	10.88
n-Heptane	6.33	o-Xylene	11.43
Isooctane	5.95		

GC Conditions

- Splitless for the first 2 minutes, split mode for the reminder of the analysis
- Injector temperature: 250°C
- Detector temperature: 310°C
- Oven: Initial temp 40°C

    Hold for 5 minutes

    Ramp at 10°C/min to 200°C

    Hold for 5 minutes (or less)

- Column: DB-1 or DB-5
- Injection volume: 1 μL
- Integrator settings: Attenuation 3

    Threshold 3

- Retention times (Table 6-4)

Equipment and Glassware

- 10-mL Teflon-septum capped vials for extracting sample charcoal
- Needle-nosed pliers for breaking the sample containers
- Capillary column gas chromatograph
- 1-, 2-, and 5-mL volumetric pipets

PROCEDURE

Week 1

1. Your instructor will assign you times and dates to sample at a local gasoline filling station. Each group will take one sample. Use a piece of plastic tubing to position the sample point at shoulder level.
2. If you are using carbon disulfide as your extraction solvent, take a sample over a 5 to 10 minute period. It typically takes 0.75 to 1.5 minutes to fill an empty tank, so you will have to take a composite sample while filling several cars. Remember to turn the pump off between cars. If you are using pentane as your extraction solvent, you will need to sample for 10 minutes.
3. Cap the ends of the sampling tube with the caps included in your kit when you are finished.

Week 2

4. Start the GC, and run your calibration standards while you prepare your samples.
5. Extract (desorb) your sample tubes as illustrated by your laboratory instructor. You will need to place the charcoal from the front and back in two separate vials.
6. Add 1.00 mL of your extraction solvent containing decane (your internal standard).
7. Cap the vial and allow it to stand for 5 minutes.
8. Analyze each sample on the GC.

Waste Disposal

All extraction solvents, calibration standards, and liquid waste should be collected in an organic waste container and disposed of by your chemistry stockroom. Your sample tubes can be disposed of in the broken-glass container.

ASSIGNMENT

1. Calculate the concentration of each analyte in an extract and the total mass of each analyte in your extraction vial.
2. Use the flow rate and sample period to convert the total mass collected to the average concentration in the air (μg/m^3 or ng/m^3).
3. Does your dose exceed the limit mentioned in the background material?

ADVANCED STUDY ASSIGNMENT

1. Draw and label a basic capillary column gas chromatograph.
2. Describe each major component in one to three sentences.

DATA COLLECTION SHEET

DATA COLLECTION SHEET

DATA COLLECTION SHEET

DATA COLLECTION SHEET

DATA COLLECTION SHEET

PART 3

EXPERIMENTS FOR WATER SAMPLES

7

DETERMINATION OF AN ION BALANCE FOR A WATER SAMPLE

Purpose: To determine the ion balance of a water sample and learn to perform the associated calculations

To learn the use of flame atomic absorption spectroscopy unit

To learn the use of an ion chromatograph unit

BACKGROUND

A favorite cartoon from my childhood shows Bugs Bunny preparing water from two flasks, one containing H^+ ions and another containing OH^- ions. Although this is correct in theory, only Bugs could have a flask containing individual ions. In reality, counterions must be present. For example, in highly acidic solutions, the H^+ ions are in high concentration but must be balanced with base ions, usually chloride, nitrate, or sulfate. In high-pH solutions, the OH^- ions are balanced by cations such as Na^+, K^+, or Ca^{2+}. The combined charge balance of the anions and cations must add up to zero in every solution. This is the principle behind the laboratory exercise presented here. You will analyze a water solution for anions by ion chromatography (IC) and for cations by flame atomic absorption spectroscopy (FAAS) and use these data to determine the ion balance of your solution. Of course, this exercise is easier than in real life, where you would have no idea which ions are present and you would have to analyze for every possible cation and anion. In this exercise we tell you which anions and cations are present.

Environmental Laboratory Exercises for Instrumental Analysis and Environmental Chemistry
By Frank M. Dunnivant
ISBN 0-471-48856-9   Copyright © 2004 John Wiley & Sons, Inc.

The presence of a variety of cations and anions in solution is very important to organisms living in or consuming the water. For example, we could not live by drinking distilled or deionized water alone. We need many of the ions in water to maintain our blood pressure and the ion balance in our cells. This need for ions in solution is important even for microorganisms living in water, since water is their medium of life. In distilled water, microbial cells try to balance the ionic strength between the internal (cell) and external water. In doing so in distilled water, the microbe cell will expand and could rupture, due to the increased volume of water required to balance the osmotic pressure across the cell membrane.

Another important point concerning ionic strength is the toxicity of inorganic pollutants, specifically metals and nonmetals. In general, the predominant toxic form of inorganic pollutants is their hydrated free ion. However, notable exceptions to this rule include organic forms of mercury and the arsenic anion. Inorganic pollutants are also less toxic in high–ionic strength (high-ion-containing) waters, due to binding and association of the pollutant with counterions in solution. This is called *complexation* and is the focus of computer models such as Mineql, Mineql+, and Geochem. For example, consider the toxicity of the cadmium metal. The most toxic form is the Cd^{2+} ion, but when this ion is dissolved in water containing chloride, a significant portion of the cadmium will be present as $CdCl^+$, a much less toxic form of cadmium. Similar relationships occur when other anions are present to associate with the free metal.

THEORY

When the concentration of all ions in solution is known, it is relatively easy to calculate an ion balance. An example is shown in Table 7-1 for a river water

TABLE 7-1. Example Calculation of the Electroneutrality of a Hypothetical River Water Sample

Ion	Molar Concentration (mol/L)	Charge Balance	Total Ion Balance
Cations			
Ca^{2+}	3.8×10^{-4}	7.6×10^{-4}	
Mg^{2+}	3.4×10^{-4}	6.8×10^{-4}	
Na^+	2.7×10^{-4}	2.7×10^{-4}	
K^+	5.9×10^{-5}	5.9×10^{-5}	
			Total cations: 1.77×10^{-3}
Anions			
HCO_3^-	9.6×10^{-4}	9.6×10^{-4}	
Cl^-	2.2×10^{-4}	2.2×10^{-4}	
F^-	5.3×10^{-6}	5.3×10^{-6}	
SO_4^{2-}	1.2×10^{-4}	2.4×10^{-4}	
NO_3^-	3.4×10^{-4}	3.4×10^{-4}	
			Total anions: 1.77×10^{-3}
			Net difference: 0.00×10^{-3}

Source: Adapted from Baird (1995).

sample. In the data analysis for this laboratory report, you must first convert from mg/L to molar concentration. Cations and anions in Table 7-1 are separated into two columns, and each molar ion concentration is multiplied by the charge on the ion. For calcium, the molar concentration of 3.8×10^{-4} is multiplied by 2 because calcium has a +2 charge. The molar charges are summarized, and if all of the predominant ions have been accounted for, the difference between the cations and anions should be small, typically less than a few percent of the total concentration. A sample calculation is included in the Advanced Study Assignment. Note that an important step in going from your analyses to your final ion balance number is to account for all dilutions that you made in the lab.

REFERENCES

Baird, C., *Environmental Chemistry*, W.H. Freeman, New York, 1995.

Berner, E. K. and R. A. Berner, *Global Environment: Water, Air, and Geochemical Cycles*, Prentice Hall, Upper Saddle River, NJ: 1996.

Dionex DX-300 Instrument Manual.

IN THE LABORATORY

Safety Precautions

- As in all laboratory exercises, safety glasses must be worn at all times.
- Use concentrated HNO_3 in the fume hood and avoid breathing its vapor. For contact, rinse your hands and/or flush your eyes for several minutes. Seek immediate medical advice for eye contact.

Glassware

- Standard laboratory glassware: class A volumetric flasks and pipets

Chemicals and Solutions

- ACS or reagent-grade NaCl, KCl, $MgSO_4$, $NaNO_3$, and $Ca(NO_3)_2$ (salts should be dried in an oven at 104°C and stored in a desiccator)
- 1% HNO_3 for making metal standards
- Deionized water
- 0.2-μm Whatman HPLC filter cartridges
- 0.2-μm nylon filters

Following are examples of preparation of IC regenerate solutions and eluents; consult the user's manual for specific compositions.

IC Regenerate Solution (0.025 N H_2SO_4). Prepare by combining 1.00 mL of concentrated H_2SO_4 with 1.00 L of deionized water. The composition of this solution will vary depending on your instrument. Consult the user's manual.

IC Eluent (1.7 mM $NaHCO_3$/1.8mM Na_2CO_3). Prepare by dissolving 1.4282 g of $NaHCO_3$ and 1.9078 g of Na_2CO_3 in 100 mL of deionized water. This 100-fold concentrated eluent solution is then diluted with 10.0 mL diluted to 1.00 L of deionized water and filtered it through a 0.2-mm Whatman nylon membrane filter, for use as the eluent. Store the concentrated solution at 4°C. Deionized water is also a reagent for washing the system after completion of the experiment. For each run, set the flow rate at 1.5 mL/min. The total cell value while running should be approximately 14 mS. Inject one or two blanks of deionized water before any standards or water samples, in order to achieve a flat baseline with a negative water peak at the beginning of the chromatogram. The composition of these solutions will vary depending on your instrument. Consult the user's manual.

IC Standards. Prepare a stock solution of the anions present in the synthetic water (chloride, nitrate, and sulfate) for each anion. For chloride, 0.208 g of NaCl should be dissolved in 100.0 mL of deionized water, yielding 1.26 g of Cl^-/L. Dilute this stock Cl^- solution 1 : 10 to give 0.126 g or 126 mg of Cl^- per litre of working standard. For nitrate, dissolve 0.155 g of $Ca(NO_3)_2$ in 100.0 mL of deionized water to yield 1172 mg NO_3^-/L. For the sulfate stock solution, dissolve 15.113 g of $MgSO_4$ in 100.0 mL of deionized water to yield 120,600 mg of SO_4^{2-}/L. An additional 1 : 1 100 mL dilution of the sulfate stock may aid in the preparation of lower-concentration sulfate standards. Thus, the working stock solution concentrations of the anions are

- 126 ppm Cl^-
- 1172 ppm NO_3^-
- 1206 ppm SO_4^{2-}

IC standards are made from the stock solutions by dilutions using 100-mL volumetric flasks and the appropriate pipets. Each calibration level shown below contains all three anions in one 100-mL volumetric flask. Final solutions should be stored in plastic bottles to prevent deterioration of the standards.

Calibration Standard I: 0.063 ppm Cl^-, 0.565 ppm NO_3^-, and 0.603 ppm SO_4^{2-}. Make a 0.05 : 100 dilution of chloride stock and nitrate stock using a 50.0- or 100.0-μL syringe and a 100 mL volumetric flask. Make a 0.05 : 100 dilution of the 1206-ppm sulfate solution using a 50.0- or 100.0-μL syringe and fill to the 100-mL mark with deionized water.

Calibration Standard II: 0.252 ppm Cl^-, 1.13 ppm NO_3^-, and 1.21 ppm SO_4^{2-}. Make a 0.2 : 100 dilution of chloride stock using a 500.0-μL syringe, a 0.1 : 100 dilution of nitrate stock using a 250.0-μL syringe, and a 0.1 : 100 dilution of the 1206-ppm sulfate solution using a 100.0-μL syringe. Fill to the 100-mL mark with deionized water.

Calibration Standard III: 1.26 ppm Cl^-, 5.65 ppm NO_3^-, and 6.03 ppm SO_4^{2-}. Make by a 1 : 100 dilution of chloride stock, a 0.5 : 100 dilution of nitrate stock using a 500.0-μL syringe, and a 0.5:100 1206-ppm sulfate solution using a 500.0-μL syringe. Fill to the 100-mL mark with deionized water.

Calibration Standard IV: 2.52 ppm Cl^-, 11.3 ppm NO_3^-, and 12.06 ppm SO_4^{2-}. Make by a 2 : 100 dilution of chloride stock, a 1:100 dilution of nitrate stock, and a 1 : 100 dilution of the 1206-ppm sulfate solution. Fill to the 100-mL mark with deionized water.

Calibration Standard V: 5.04 ppm Cl^-, 22.6 ppm NO_3^-, and 24.12 ppm SO_4^{2-}. Make by a 4 : 100 dilution of chloride stock, a 2 : 100 dilution of nitrate stock, and a 2 : 100 dilution of the 1206-ppm sulfate solution. Fill to the 100-mL mark with deionized water.

Calibration Standard VI: 11.34 ppm Cl^-, 50.85 ppm NO_3^-, and 54.45 ppm SO_4^{2-}. Make by a 1 : 10 dilution of chloride stock, a 0.5 : 10 dilution of nitrate stock using a 500.0-μL syringe and a 0.5 : 10 dilution of the 1206-ppm sulfate solution using a 500.0-μL syringe. Fill to the 100-mL mark with deionized water.

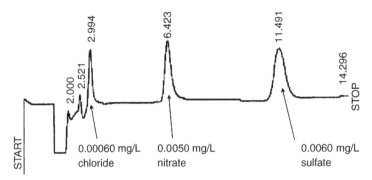

Figure 7-1. IC output for chloride, nitrate, and sulftate.

Each calibration standard solution should be filtered through a 0.45-μm Whatman HPLC filter cartridge and injected into the ion chromatograph system twice. Average peak areas should be taken based on the two injections and used to produce linear calibration graphs using the linear least squares Excel program described in Chapter 2.

To aid in your analysis, a typical ion chromatogram of chloride, nitrate, and sulfate is shown in Figure 7-1. Your retention times may differ from those shown below, but the elution order should be the same. Adjust the elution times to have a total run time of less than 15 minutes.

FAAS Standards. The cations in the synthetic water are Ca^{2+}, Mg^{2+}, Na^+, and K^+. Unlike the IC solution preparation, you must figure out how to make the calibration solutions. Stock solution concentrations should be 1000 ppm (mg/L) for each cation made from the dried and desiccated salts. Standards should be made for each cation using the approximate solution concentrations shown in the list that follows. Note that you will have to make serial dilutions of the 1000-mg/L stock solution to obtain the concentration shown below using standard class A pipets. The exact range and approximate concentrations of standards and detection limits may vary depending on the FAAS unit that you use. You may have to lower or raise the standard concentrations.

- Ca^{2+}: 1 ppm, 5 ppm, 10 ppm, 15 ppm, 20 ppm, 25 ppm, and 50 ppm
- Mg^{2+}: 0.05 ppm, 0.1 ppm, 0.2 ppm, 0.5 ppm, 1 ppm, 1.5 ppm, and 2 ppm
- Na^+: 0.2 ppm, 0.5 ppm, 1 ppm, 3 ppm, 5 ppm, 10 ppm, and 12 ppm
- K^+: 0.5 ppm, 1 ppm, 2 ppm, 3 ppm, 4 ppm, and 5 ppm

Each element will be analyzed using FAAS to create a linear calibration curve for each cation. The data can be analyzed using the linear least squares Excel sheet described in Chapter 2. You will be given a water sample by your instructor that contains each of the cations and anions mentioned above. You must determine the concentrations of each ion. Alternatively, the cations can be analyzed by IC. Consult the user's manual for specific details.

PROCEDURE

Limits of the Method. (These will vary depending on the instrument you use.)

Anions

- 0.0001 ppm Cl^-
- 0.01 ppm SO_4^{2-}
- 0.002 ppm NO_3^-

Cations

- 0.4 ppm Ca^{2+}
- 0.02 ppm Mg^{2+}
- 0.002 ppm Na^+
- 0.1 ppm K^+

This laboratory exercise will take three 4-hour laboratory periods if you are asked to perform all experiments. Alternatively, your professor may divide you into three groups: an IC group, a Ca and Mg group, and a Na and K group. If you are divided into groups, the entire exercise can be completed in one lab period, but you will be sharing your results with the remainder of the class.

IC Analysis

1. First, sign in the logbook, turn on the IC, and start the system. This will allow the eluent, column, and detector to equilibrate while you prepare your calibration standards.
2. Prepare your calibration standards as described above.
3. Dilute your water sample 1 : 500, 1 : 250, 1 : 100, and 1 : 1 for analysis, and in step 4, analyze each sample from low to high concentration until you determine the appropriate dilution to be analyzed. Analyze each water sample twice as time permits, and determine the most appropriate sample dilution based on your calibration curve (again from step 4).
4. Analyze your IC standards and then your unknown samples, making duplicate injections as time permits. Remember to record any instrument problems in the logbook as you sign out.
5. Use the linear least squares Excel program to analyze your data.

FAAS Analysis

1. First, turn on the FAAS unit and lamp. This will allow the system to warm up while you prepare your calibration standards and sample dilutions.

2. Note that all solutions/dilutions should be made in 1% HNO_3 to preserve your samples and standards.

3. Prepare your FAAS calibration standards as described earlier.

4. Dilute your water sample 1 : 500, 1 : 250, 1 : 100, and 1 : 1 and analyze each sample from low to high concentration until you determine which dilution is appropriate for analysis. Analyze each sample twice as time permits and determine the most appropriate dilution based on your calibration curve.

5. Analyze each metal separately.

6. Use the linear least squares Excel program to analyze your data.

Waste Disposal

After neutralization, all solutions can be disposed of down the drain with water.

ASSIGNMENT

Calculate the ion balance for your water sample **based on the undiluted solution**.

ADVANCED STUDY ASSIGNMENT

1. Why is the electroneutrality of a water sample important to document?
2. How do the anion and cation content affect toxicity?
3. Using your library's online search engine, find an example in the literature describing the toxicity of a complexed metal ion. The two important journals *Environmental Science and Technology* and *Environmental Toxicology and Chemistry Journal* should be included in your search.
4. Complete Table 7-2 to determine the net electroneutrality of the water sample. Is the solution balanced with respect to cations and anions?

TABLE 7-2. Calculation of the Electroneutrality of Seawater

Ion	Concentration (mg/L)	Molar Concentration (mol/L)	Charge Balance	Total Ion Balance
Cations				
Ca^{2+}	4,208			
Mg^{2+}	1,320			
Na^+	11,012			
K^+	407			
				Total cations:
Anions				
HCO_3^-	122			
Cl^-	19,780			
SO_4^{2-}	2,776			
				Total anions:
				Net difference:

Source: Based on data in Berner and Berner (1996).

DATA COLLECTION SHEET

DATA COLLECTION SHEET

DATA COLLECTION SHEET

DATA COLLECTION SHEET

8

MEASURING THE CONCENTRATION OF CHLORINATED PESTICIDES IN WATER SAMPLES

Purpose: To determine the concentration of chlorinated pesticides in a water sample

To use a capillary column gas chromatograph equipped with an electron-capture detector

BACKGROUND

Chlorinated pesticides are considered to be ubiquitous in the environment due to their refractory behavior (very slow chemical and biochemical degradation) and widespread use. For example, chemicals such as DDT and PCBs have been observed in water, soil, ocean, and sediment samples from around the world. Although the production and use of these chemicals has been banned in the United States since the 1970s, many countries (with the help of American-owned companies) continue to produce and use these chemicals on a routine basis.

Chlorinated hydrocarbons can be detected at incredibly low concentrations by a gas chromatograph detector [the electron-capture detector (ECD)] developed by James Lovelock (also the originator of the Gaia hypothesis, described in the background section of Chapter 5.) In fact, the first version of this detector was so sensitive that the company reviewing Lovelock's proposal did not believe his results and rejected his findings. Lovelock persisted and today is responsible for one of the most important and most sensitive GC detectors. The ECD can detect less than a picogram of a chlorinated compound. But with this sensitive detection

Environmental Laboratory Exercises for Instrumental Analysis and Environmental Chemistry
By Frank M. Dunnivant
ISBN 0-471-48856-9   Copyright © 2004 John Wiley & Sons, Inc.

limit comes a dilemma: How sensitive should our environmental monitoring be? Although the wisdom behind this policy is questionable, we set many exposure limits for pesticides based on how little of it we can measure with our expensive instruments. As we develop better and better instruments, we push the detection limits lower, and consequently, we set our exposure limits lower. Given the long-term presence of these compounds, we seem to be chasing a never-ending lowering of the exposure limits. Thus, we often turn to toxicology studies to determine exactly what level of exposure is acceptable.

The determination of the solubility of a specific compound is a relatively straightforward process in pure distilled water, and solubility values can be found in the literature. But how relevant are these published values to real-world samples? Literature values are available for the maximum solubility of compounds in water. In general, solubilities of hydrophobic compounds increase with temperature. But if you take a lake water sample and measure the concentration of DDT, is the DDT present only in the dissolved phase? One highly complicating factor in solubility measurements is the presence of a "second phase" in natural water samples that is usually described as colloidal in nature. Colloids can take the form of inorganic particles that are too small to filter from the sample or as natural organic matter (NOM) that is present in most water samples. Hydrophobic pollutants in water greatly partition to these additional particles in water and result in an apparent increase in water solubility. So if you measure the pesticide concentration of a water sample and your data indicate that you are above the water solubility, the solution may not actually be supersaturated but rather, may contain a second phase that contains additional analyte. Scientists have developed ways to detect the presence of colloid and colloid-bound pollutants, but these techniques are beyond the scope of this manual.

In this laboratory experiment you will be using a separatory funnel extraction procedure to measure the concentration of chlorinated pesticides in a water sample. This water sample is relatively pure and does not contain appreciable amounts of a second phase. This technique has been used for decades to monitor the presence of pesticides in water samples.

THEORY

If you consider only one contact time in the separatory funnel, we can define a distribution ratio, D, which describes the equilibrium analyte concentration, $C_{organic}$, between the methylene chloride and the water, C_{water}, phases:

$$D = \frac{[C]_{\text{methylene chloride}}}{[C]_{\text{water}}}$$

The extraction efficiency is given by

$$E = \frac{100D}{D + V_{\text{methylene chloride}}/V_{\text{water}}}$$

When D is greater than 100, which it is for most hydrophobic analytes, a single equilibrium extraction will quantitatively extract virtually all of the analyte into the methylene chloride phase. However, as you will note during the experiment, some of the methylene chloride will stick to the sides of the separatory funnel and not pass into the collection flask (a 100-mL volumetric flask). To achieve complete recovery of the methylene chloride, as well as complete extraction, you will extract the sample three times and combine the extractions in a 100-mL collection flask.

We can also estimate how many extractions are necessary to remove a specified quantity of the analyte for a series of extractions. This effectiveness can be evaluated by having an estimate of D and calculating the amount of solute remaining in the aqueous phase, $[C]_{water}$, after n extractions, where

$$[C_{water}]_n = C_{water}\left[\frac{V_{water}}{DV_{organic} + V_{water}}\right]^n$$

ACKNOWLEDGMENT

I would like to thank Josh Wnuk for the experimental design, data collection, and analysis.

REFERENCES

Fifield, F. W. and P. J. Haines, *Environmental Analytical Chemistry*, 2nd ed., Blackwell Science, London, 2000.

Perez-Bendito, D. and S. Rubio, *Environmental Analytical Chemistry*, Elsevier, New York, 2001.

IN THE LABORATORY

Your laboratory procedure involves the extraction of very low concentrations of chlorinated pesticide/PCB in water. You will accomplish this by performing three extractions in a separatory funnel, combining these extracts, and concentrating the extract for analysis on a GC. Finally, you will analyze your samples on a capillary column GC equipped with an electron-capture detector.

Safety Precautions

- Safety glasses must be worn at all times during this laboratory experiment.
- Most, if not all of the compounds that you will use are carcinogens. Your instructor will prepare the aqueous solution of these compounds so that you will not be handling high concentrations. The purge solution you will be given contains ppb levels and is relatively safe. You should still use caution when using these solutions since the pesticides and PCBs are very volatile when placed in water. Avoid breathing the vapors from this solution.
- Most of the solvents used in this experiment are flammable. Avoid their use near open flames.

Chemicals and Solutions

Neat solutions of the following compounds will be used by your instructor to prepare the aqueous solution:

- Lindane
- Aldrin
- 2,2',4,4',6,6'-Hexachlorobiphenyl
- Dieldrin (not added to the solution to be extracted, but to be used as a analyte recovery check standard)
- Endosulfan I (not added to the purge solution, but to be used as a GC internal standard)

You will need, in addition:

- 80.0-ppm solution of Endosulfan I
- 80.0-ppm solution of Dieldrin
- Solid NaCl (ACS grade)
- Anhydrous Na_2SO_4 dried at $104°C$

Glassware

For each student group:

- 1-L separatory funnel
- ~10 cm by ~2.0 cm drying column
- 100.0-mL volumetric flask
- Pasteur pipets
- Two 5- or 10-µL microsyringes

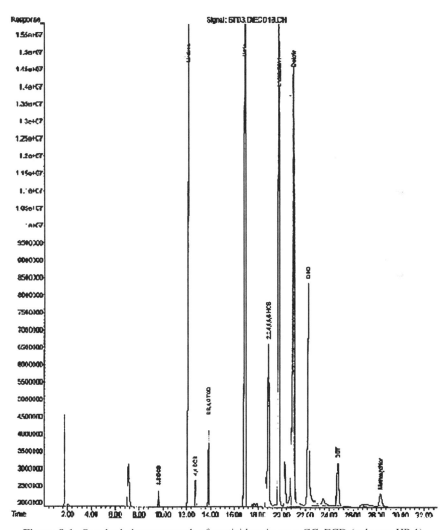

Figure 8-1. Standard chromatograph of pesticide mix on a GC–ECD (column: HP-1).

GC Conditions

- 1.0-μL injection
- Inlet temperature = 270°C
- Column:
    HP-1 (cross-linked methyl silicone gum)
    30.0 m (length) by 530 μm (diameter) by 2.65 μm (film thickness)
    4.02-psi column backpressure
    3.0-mL/min He flow
    31-cm/s average linear velocity
- Oven:
    Hold at 180°C for 1.0 minute
    Ramp at 5.0°C/min
    Hold at 265°C for 16.0 minutes
    Total time = 34.0 minutes
- Detector:
    Electron-capture detector
    Temperature = 275°C
    Makeup gas = $ArCH_4$
    Total flow = 60 mL/min
- Retention times (from Figure 8-1) for the given GC setting are:

Lindane	12.13 minutes
Aldrin	16.86 minutes
2,2′,4,4′,6,6′-TCB	18.86 minutes
Endosulfan I (IS)	19.75 minutes
Dieldrin	20.95 minutes

PROCEDURE

1. Obtain a water sample from your laboratory instructor. The water sample will be a 500- or 1000-mL glass bottle and will contain a known concentration of each analyte.

2. Set up your extraction apparatus according to Figure 8-2. Soap-wash and water-rinse all glassware that will be contacting your sample to remove interfering compounds (especially phthalates from plastics). Remove any water with a minimal amount of pesticide-grade methanol or acetone. Finally, rinse the glassware with pesticide-grade methylene chloride. Deposit rinse solvents in an organic waste bottle, *not* down the sink drain.

3. Fill the drying column with anhydrous Na_2SO_4 (a 3- to 4-inch column of Na_2SO_4 will be sufficient).

4. Pour the contents of your sample container into your separatory funnel. Add about 25 mL of methylene chloride to your original sample container, cap it, and shake for 30 seconds. (The purpose of this step is to remove any analyte that may have sorbed to the surface of your sample container.)

5. Quantitatively transfer the methylene chloride from your sample container to the separatory funnel. Add about 1 g of NaCl to your water sample in the separatory funnel (this will inhibit the formation of an emulsion layer that could form between the two liquid layers and interfere with your transfer to the drying column). Seal the funnel, shake vigorously for 2 minutes, releasing the pressure as necessary, and allow the layers to separate. Swirl the funnel as needed to enhance the separation and remove methylene chloride from the separatory funnel walls.

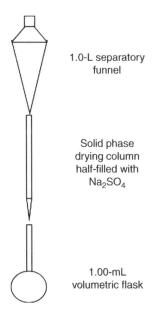

1.0-L separatory
funnel

Solid phase
drying column
half-filled with
Na_2SO_4

1.00-mL
volumetric flask

Figure 8-2. Extraction setup held in place with a ring stand.

6. Carefully open the stopcock and allow only the bottom layer (methylene chloride) to enter the drying column. Be careful not to let any water phase enter the drying column since excessive amounts of water will clog this column. The methylene chloride should pass uninhibited into the 100.0-mL volumetric flask.

7. Add about 25 mL of methylene chloride to your sample container and repeat steps 4 through 6 two more times, collecting each extract into the 100.0-mL volumetric flask. (As you add methylene chloride to the drying column, you may occasionally need to break up the surface of the column. Water contained in the methylene chloride will be removed from the organic layer and bound to the Na_2SO_4, forming a crust on the surface.)

8. Rinse the drying column with additional methylene chloride and fill your 100-mL volumetric flask to the mark.

9. The concentration in your water sample and methylene chloride extract is very low and needs to be concentrated to measure the concentration adequately. We will concentrate your extract using a warm water bath and a gentle flow of N_2 (or He). Pipet 10.00 mL of your 100.0-mL extract into a graduated 10- or 15-mL thimble. We will check the recovery of this step using an internal standard, Dieldrin. Using a microsyringe, add exactly 2.00 μL of an 80.0-ppm Dieldrin solution supplied by your laboratory instructor. Place the thimble in a warm water bath and adjust a gentle stream of nitrogen or helium over the surface of the liquid. The gas stream will evaporate the liquid.

10. After the liquid level has reached ~1 mL, pipet 5.00 mL more of your extract into the thimble (this will give you a total of 15.0 mL). Gently evaporate the liquid to dryness, remove immediately, and add isooctane and your GC internal standard. First, pipet 2.00 mL of isooctane into the thimble. The GC internal standard is Endosulfan I. Using another microsyringe, add 2.00 μL of an 80.0-ppm solution. Using a clean Pasteur pipet, rinse the walls of the thimble from top to bottom several times. This will redissolve any analyte or internal standards that precipitated on the walls of your thimble. The final concentration of each internal standard is 32.0 ppb.

11. Transfer the isooctane extract to a GC autoinjection vial or cap your thimble until you analyze it on the GC.

12. Sign into the GC logbook and analyze your samples using the conditions given under "GC Conditions" in the section "In the Laboratory." When you finish, record any instrument problems in the logbook and sign out.

Waste Disposal

All organic liquids should be disposed of in an organic hazardous waste receptacle. These solutions will be disposed of properly by the safety officer.

ASSIGNMENT

After you analyze your samples, calculate the concentration of each analyte in your original water sample. Calculate a standard deviation using data acquired by the entire class. Using the Student t-test spreadsheet (see Chapter 2) and the known value provided by your instructor, determine if bias is present in your analysis.

ADVANCED STUDY ASSIGNMENT

A water sample is extracted for DDT and analyzed by GC–ECD. A 500-mL water sample is extracted three times using a separatory funnel and the extract is combined to a final volume of 100.0 mL. A 20.00-mL aliquot of the 100.0 mL is concentrated to 1.00 mL. Dieldrin is added as a recovery check standard to the 1.00-mL concentrated extract at a concentration of 50.0 ppb. A GC internal standard is added to correct for injection errors and is recovered at 95.0%. Calculate the concentration of DDT in your original water sample using the following data:

- *GC results for DDT:* 45.6 µg/L in the 1.00-mL concentrated solution
- *GC results for Dieldrin:* 48.5 µg/L

DATA COLLECTION SHEET

DATA COLLECTION SHEET

DATA COLLECTION SHEET

DATA COLLECTION SHEET

9

DETERMINATION OF CHLORIDE, BROMIDE, AND FLUORIDE IN WATER SAMPLES

Purpose: To learn to use ion-specific electrodes

To determine the concentration simple anions in water samples

BACKGROUND

As rainwater falls on the Earth and contacts soil, it dissolves minerals, which are washed into streams and lakes. These waters, in turn, transport a variety of cations and anions to the oceans. Over millions of years, this resulted in the high salt content of ocean water. Common cations include sodium, potassium, calcium, and magnesium; common anions are chloride, sulfate, carbonate, bicarbonate, and nitrate, although other cations and anions may be present, depending on the local geologic media. Some ions are nutrients; others may be potentially toxic. In this laboratory we use a relatively simple method for measuring the activity of anions in water. Note that electrodes measure activity, not concentration. In low–ionic strength waters, the activity is essentially equal to concentration, but for higher ionic strengths, important differences in these measurements are present.

THEORY

Ion-specific electrodes are a convenient and easy way to determine the concentration of certain ions in solution. A variety of electrode designs are available,

Environmental Laboratory Exercises for Instrumental Analysis and Environmental Chemistry
By Frank M. Dunnivant
ISBN 0-471-48856-9   Copyright © 2004 John Wiley & Sons, Inc.

including (1) liquid membrane electrodes that measure Ca^{2+}, BF_4^-, NO_3^-, ClO_4^-, K^+, Ca^{2+}, and Mg^{2+} (water hardness); (2) gas-sensing probes that measure NH_3, CO_2, HCN, HF, H_2S, SO_2, and NO_2; and (3) crystalline membrane electrodes (solid-state electrodes) that measure Br^-, Cd^{2+}, Cl^-, Cu^{2+}, F^-, I^-, Pb^{2+}, Ag/S^{2-}, and SCN^-. We use the latter, solid-state electrodes to measure Cl^-, Br^-, and F^- ion concentrations.

The operation of solid-state electrodes is similar to that of the glass, pH electrode. A potential is established across a membrane. In a pH electrode, the membrane is a semipermeable glass interface between the solution and the inside of the electrode, while in solid-state electrodes, the membrane is a 1- to 2-mm-thick crystal. For example, for the fluoride electrode, the crystal is composed of lanthanum fluoride (LaF_3) doped with europium fluoride (EuF_2). At the two interfaces of the membrane, ionization occurs and a charge is created described by

$$LaF_3(s) \leftrightarrow LaF_2^+(s) + F^-(aq)$$

The magnitude of this charge is dependent on the fluoride ion concentration in the test sample or standard. A positive charge is present on the side of the membrane that is in contact with the lower fluoride ion concentration, while the other side of the membrane has a negative charge. The difference in charge across the membrane allows a measure of the difference in fluoride concentration between the two solutions (inside the electrode and in the test solution).

The solid-state electrodes are governed by a form of the *Nernst equation*,

$$E = K + \frac{0.0592}{n} pX \tag{9-1}$$

where E is the voltage reading, K an empirical constant (the y intercept of the log-activity or concentration plot), $0.0592/n$ the slope of the line [$0.0592 = RT/F$ ($R = 8.316$ J/mol·K, T in temperature in kelvin, and $F = 96487$ C/mol)], and pX is the negative log of the molar ion concentration. Note that for monovalent ions (an n value of 1), the slope should be equal to 0.0592 if the electrode is working properly. If a significantly different slope is obtained, the internal and external filling solutions of the reference electrode should be changed, or the end of the solid-state electrode should be cleaned.

You should note that the semipermeable membrane provides only one-half of the necessary system, and a reference electrode is needed. There are three basic types of reference electrodes: the standard hydrogen electrode, the calomel electrode, and the Ag/AgCl electrode. Most chemists today use the Ag/AgCl reference electrode. This addition gives us a complete electrochemical cell. Note that a plot of the log of ion activity versus the millivolt response must be plotted to obtain a linear data plot. Also note that the concentration can be plotted as log(molar activity) or log(mg/L).

REFERENCES

Skoog, D. A, F. J. Holler, and T. A. Nieman, *Principles of Instrumental Analysis*, Saunders College Publishing/Harbrace College Publishers, Philadelphia, 1998.

Willard, H. H., L. L. Merritt, Jr., J. A. Dean, and F. A. Settle, Jr., *Instrumental Methods of Analysis*, Wadsworth, Belmont, CA, 1988.

IN THE LABORATORY

Safety Precautions

- Safety glasses should be worn at all times during the laboratory exercise.
- The chemicals used in this laboratory exercise are not hazardous, but as in any laboratory, you should use caution.

Chemicals

- Sodium or potassium salts of chloride, bromide, or fluoride (depending on the ion you will be analyzing)
- Ionic strength adjustor (consult the user's manual)

Equipment and Glassware

- Solid-state electrodes (each ion will have a specific electrode)
- Ag/AgCl reference electrode
- mV meter
- Standard volumetric flasks
- Standard beakers and pipets

PROCEDURE

The exact procedure will depend on the brand of electrode you are using. Consult the user's manual. In general, you will need an ionic strength adjustor that does not contain your ion of interest, a single- or double-junction reference electrode (specified in the solid-state electrode user's manual), and a set of reference standards made from the sodium or potassium salts. In general, the range of standards should be from 0.50 to 100 mg/L.

1. First, set up your electrodes and allow them to equilibrate in the solution for the time specified in the user's manual.
2. Make up your reference standards and analyze them from low to high concentration.
3. Make a plot according to equation (9-1) (mV versus the negative log of your analyte concentration) and ensure that the slope is at or near 59.2.
4. Analyze your unknown samples.
5. Calculate the concentration in your samples.
6. Disassemble the setup. Dry off the solid-state electrode and return it to its box. Empty the filling solution of the reference electrode, wash the outside and inside with deionized water, and allow it to air dry.

Waste Disposal

All solutions can be disposed of down the drain with excess water.

ASSIGNMENT

Use the Excel spreadsheet to analyze your data. Calculate the concentration of analytes in your samples.

ADVANCED STUDY ASSIGNMENT

Research solid-state electrodes. Draw a complete electrode setup, including a cross section of a solid-state electrode and a cross section of an Ag/AgCl reference electrode.

DATA COLLECTION SHEET

DATA COLLECTION SHEET

DATA COLLECTION SHEET

DATA COLLECTION SHEET

DATA COLLECTION SHEET

10

ANALYSIS OF NICKEL SOLUTIONS BY ULTRAVIOLET–VISIBLE SPECTROMETRY

SAMANTHA SAALFIELD

Purpose: To determine the concentration of a transition metal in a clean aqueous solution

To gain familiarity with the operation and applications of an ultraviolet–visible spectrometer

BACKGROUND

When electromagnetic radiation is shown through a chemical solution or liquid analyte, the analyte absorbs specific wavelengths, corresponding to the energy transitions experienced by the analyte's atomic or molecular valence electrons. Ultraviolet–visible (UV–Vis) spectroscopy, which measures the absorbent behavior of liquid analytes, has in the last 35 years become an important method for studying the composition of solutions in many chemical, biological, and clinical contexts (Knowles and Burges, 1984).

UV–Vis spectrometers operate by passing selected wavelengths of light through a sample. The wavelengths selected are taken from a beam of white light that has been separated by a diffraction grating. Detectors (photomultiplier tubes or diode arrays) report the amount of radiation (at each wavelength) transmitted through the sample. The peaks and troughs of absorption at different wavelengths for a particular analyte are characteristic of the chemicals present, and the concentration of chemicals in the sample determines the amount of

Environmental Laboratory Exercises for Instrumental Analysis and Environmental Chemistry
By Frank M. Dunnivant
ISBN 0-471-48856-9   Copyright © 2004 John Wiley & Sons, Inc.

radiation reaching the detector. Thus, for a given solution, the wavelength of maximum absorption (λ_{max}) remains constant, while the percent transmittance increases and the absorbance decreases as the solution is diluted (as will be seen in this experiment).

Major limitations of UV–Vis spectroscopy result from the nonspecific nature of the instrument. Spectrometers simply record how much radiation is absorbed, without indicating which chemical species is (are) responsible. Thus, spectroscopy is most valuable in analyzing clean solutions of a single known species (often at different concentrations, as studied in this experiment), or analytes such as plating solutions, which have only one (metal) species that will absorb visible light. Procedures for activating a particular species, or giving it color through chemical reaction, can also make spectroscopy useful for analyzing complex matrices.

UV–Vis spectroscopy has various applications in environmental chemistry. For plating solutions, knowing the amount of metal present in waste determines treatment procedures. Complex extraction and digestion procedures are also used to determine the concentrations of species, from iron to phosphate, in soils, sediments, and other environmental media.

THEORY

The relationship between absorbance and concentration for a solution is expressed by *Beer's law*:

$$A = \varepsilon bc = -\log T \tag{10-1}$$

where A is the absorbance by an absorbing species, ε the molar absorptivity of the solution, independent of concentration ($\text{L/mol} \cdot \text{cm}$), b the path length of radiation through cell containing solution (cm), and c the concentration of the absorbing species (mol/L). Thus, when the molar absorptivity (dependent on the atomic or molecular structure) and path length are held constant, the absorbance by an analyte should be directly proportional to the concentration of the absorbing species in the analyte. This leads to a linear relationship between concentration and absorbance and allows the concentration for unknown samples to be calculated based on plots of data for standards of known concentrations. If more than one absorbing species is present, the absorbance should be the sum of the absorbances of each species, assuming that there is no interaction between species.

Beer's law generally holds true for dilute solutions (where absorbance is less than 3). At higher concentrations, around the *limit of quantitation*, the plot of concentration versus absorbance levels out. This occurs as the absorbing species interferes with itself so that it can no longer absorb at a rate proportional to its concentration. A leveling out of the Beer's law plot may also be observed at very low concentrations, approaching the *limit of linearity* and the detection limit of the instrument.

The absorbance of electromagnetic radiation by chemical compounds in solution results from the transitions experienced by the compounds' electrons in response to the input of photons of distinct wavelengths. Organic compounds often contain complex systems of bonding and nonbonding electrons, most of which absorb in the vacuum–UV range (less than 185 nm). Functional groups that allow excitation by, and absorbance of, radiation in the longer UV or visible wavelengths are called *chromophores*. For example, unsaturated functional groups, containing nonbonding (*n*) or pi-orbital (π) electrons, absorb between 200 and 700 nm (often in the visible range) as they are excited into the antibonding pi orbital (π^*).

The absorption of visible radiation by light transition metals leads to primary applications of spectroscopy to inorganic compounds. These metals have a characteristic set of five partially filled *d* orbitals, which have slightly different energies when the metals are complexed in solution. This enables electronic transitions from *d* orbitals of lower to higher energies. In solutions of divalent metals with nitrate, such as the solution of $Ni(NO_3)_2 \cdot 6H_2O$ that we study in this experiment, six water molecules generally surround the dissolved metal in an octahedral pattern (Figure 10-1). The negative ends of these molecules, aligned toward the unfilled *d* orbitals of the metal, repel the orbitals and thus increase their energy. However, due to the distinct orientations of the various *d* orbitals around the nucleus, some are more affected than others by this repulsion. The relatively small resulting energy differences correspond to photons in the visible range. For lightweight transition metals, these wavelengths vary according to the solvent (in this experiment, water) and resulting ligand ($Ni(H_2O)_6^2$; in contrast, the spectra for lanthanide and actinide metals have sharper peaks and are generally independent of solvent. Overall, the subtle *d*-orbital splitting in transition metal solutions gives these solutions their colors and makes them valuable candidates for visible spectrometric analysis.

Although all spectrophotometers operate on the same principles, they have a number of variations that affect their operation and analytical flexibility. Some instruments have adjustable bandwidths, which allow you to change the amount of

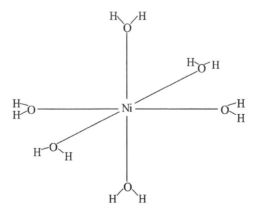

Figure 10-1. Model of octahedral nickel ion–water complex.

the diffracted light that the instrument allows through to the sample. Narrow slit widths allow a finer resolution, while widening the bandwidth gives a stronger signal. One consideration regarding both bandwidths and analyte concentrations is the signal-to-noise ratio of the results. Like all instruments, spectrophotometers have some background signal, a "noise" that is manifested as the standard deviation of numerous replicate measurements. With either narrow slit widths or lower concentrations, the *signal-to-noise ratio* (average reading/standard deviation) may increase due to a decrease in the signal, although this is more significant in regard to concentrations.

Spectrophotometers may also be single- or double-beam, the primary difference being the continual presence of a blank cell in the double beam, eliminating the need for repeated reference measurements, since during each measurement the beam of radiation passing through the analyte cell also passes through the reference cell on its way to the detector. Also, whereas in older, nonautomated spectrophotometers it was preferable to take measurements of percent transmittance because they gave a linear plot, on newer digital machines it is fine to read absorbance directly.

REFERENCES

Knowles, A. and C. Burges (eds.), *Practical Absorption Spectrometry*, Vol. 3, Chapman & Hall, London, 1984.

Sawyer, D. T., W. R. Heineman, and J. M. Beebe, *Chemistry Experiments for Instrumental Methods*, Wiley, New York, 1984.

Skoog, D. A., J. F. Holler, and T. A. Nieman, *Principles of Instrumental Analysis*, 5th ed., Harcourt Brace College Publishing, Philadelphia, 1998.

IN THE LABORATORY

Chemicals

- ACS-grade crystalline $Ni(NO_3)_2 \cdot 6H_2O$

Equipment and Glassware

- Spectrophotometer (automated is preferable, but a Spectronics 20 will work), with visible radiation lamps
- Analytical balance
- Five 25-ml volumetric flasks per student or pair of students
- 1-mL, 2-mL, 4-mL, and 10-mL pipets
- Matched cuvettes for visible light

Preparation of Standards

- $0.250\,M$ $Ni(NO_3)_2 \cdot 6H_2O$: Dissolve about 1.82 g of crystalline $Ni(NO_3)_2 \cdot 6H_2O$ in deionized water in a 25-mL volumetric flask. Record the actual weight of $Ni(NO_3)_2 \cdot 6H_2O$ added, to calculate the actual concentration.
- *Dilutions*: $0.0100\,M$, $0.0200\,M$, $0.0400\,M$, and $0.100\,M$ $Ni(NO_3)_2 \cdot 6H_2O$: Pipet 1.00 mL, 2.00 mL, 4.00 mL, and 10.00 mL of $0.250\,M$ $Ni(NO_3)_2 \cdot 6H_2O$, respectively, into 25-mL volumetric flasks. These and the remaining $0.250\,M$ solution can be stored in covered beakers if necessary or to make them easier to transfer.

PROCEDURE

1. Turn on the spectrophotometer and allow it to warm up for 20 minutes.

2. If the spectrophotometer is connected to a computer, turn the computer on and open the appropriate program.

3. Use the $0.100 M$ $Ni(NO_3)_2 \cdot 6H_2O$ solution to test for maximum absorbance (λ_{max}). Rinse the cuvette with deionized water, followed by a small portion of the analyte solution, and then pour about 3 mL of solution into a cuvette. Zero the spectrophotometer. If your instrument will scan across a range of wavelengths, perform a scan from 350 to 700 nm. If not, you need to test the absorbance of the solution every 5 nm across this range. Record the location of the largest, sharpest peak. Retain the cuvette with $0.100 M$ nickel for use in step 5.

4. If working on a computer, open the fixed-wavelength function. Set the wavelength to the λ_{max} you found in step 3 on either the computer or the manual dial. If bandwidth is adjustable, set it at 2 nm. Rezero the instrument.

5. Analyze the $0.100 M$ nickel solution already in the cuvette at λ_{max}. Repeat 5 to 10 times, and record the absorbance readings. Empty the cuvette, rinse it with deionized water and with the $0.0100 M$ solution, fill it with the $0.0100 M$ solution, and analyze the contents 5 to 10 times. Repeat this process for each of the remaining three solution concentrations, proceeding from least to most concentrated.

6. Obtain an unknown in a 25-mL volumetric. Determine it absorbance at λ_{max}, taking five measurements.

Note on blank measurements: If you are using an automatic spectrophotometer, you only need to take blank measurements at the beginning and end of the day. If you are on a manual instrument, take blank measurements often, such as when you change solutions or parameters of measurements.

Optional Procedures

Signal-to-Noise Ratio

1. Analyze three or more of the nickel concentrations at least 20 times, recording each absorbance, and calculate the mean and standard deviation about the mean of the repetitive measurements. (signal-to-noise ratio = mean/standard deviation).

2. Compare the signal-to-noise ratios for the various concentrations. What effect does changing concentration have on the ratio? What implication does that have for the quality of results?

Wavelength and Signal-to-Noise Ratio

1. Analyze one or more of the nickel concentrations at more than one wavelength (λ_{max} and at least one at nonpeak absorbance) with at least 20 repetitions for each wavelength. Be sure to rezero the instrument each time you change the wavelength.
2. Compare the absorbance at various wavelengths. Does the trend make sense? Compare the signal-to-noise ratios for the same concentration at different wavelengths. What effect does changing wavelength have on the ratio? What implication does that have for the quality of results?

Slit Width and Signal-to-Noise Ratio. This requires an instrument with adjustable bandwidths.

1. Analyze two or more of the nickel concentrations at multiple bandwidths (e.g., 0.5 nm, 2 nm, 10 nm), with at least 20 repetitions for each bandwidth. Be sure to rezero the instrument each time you change the bandwidth.
2. Compare the absorbances and the signal-to-noise ratios for various bandwidths.

Note: To conserve solutions in carrying out these optional procedures, work with one solution at a time by incorporating these procedures into step 5 of the main procedure. [The frequent changing of settings (precedents) that this requires may make it difficult on a Spectronics 20 (nonautomated) system.] For example, if you plan to complete all the procedures, when you get to step 5, scan the 0.100 M solution 20 times (at $\lambda = \lambda_{max}$ and bandwidth = 2 nm). Then change the wavelength and scan 20 times again. Return the wavelength to λ_{max}, change the bandwidth, and scan at 0.5 nm and then at 10 nm. Restore the original settings and proceed to the other solutions, carrying out as many of the optional procedures as desired. The *most important* thing to remember is to rezero the instrument each time you change the wavelength or bandwidth.

Waste Disposal

Nickel solutions should be placed in a metal waste container for appropriate disposal.

ASSIGNMENT

1. Create a Beer's law plot similar to the one shown in Figure 10-2, relating nickel concentration (*x* axis) to mean absorbance (*y* axis) for the standard solutions. Be sure to use the actual concentrations of the solutions you made if they varied from the stated value. Turn in a copy of this plot along with a short table of the corresponding data (mean absorbances and concentrations).

2. Complete a linear least squares analysis on the Beer's law plot, using the statistical template spreadsheet provided on the included CD-ROM or from your instructor. Turn in a copy of the spreadsheet with a short discussion of what the analysis indicates about your data.

3. Evaluate your unknowns. After you have entered the data for the standards into the "LLS" spreadsheet, enter the absorbances ("signals") of the unknowns into the bottom of the sheet. Transfer the concentrations calculated by Excel for these absorbances into the "*t*-test" sheet ("observation" column). Enter the number of replicates (*N*), and set the desired degrees of freedom (usually, *N* − 1) and the confidence interval. Fill in the true unknown concentrations provided by your instructor, and consult the statistical test to see whether bias is present in your measurements. Include a copy of the spreadsheets in your lab manual with a short discussion of what this test indicates and of possible sources of discrepancy between your calculated concentration values and the true values.

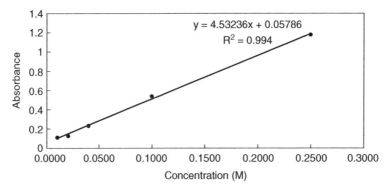

Figure 10-2. Example of typical student data: Beer's law plot for $Ni(NO_3)_2 \cdot 6H_2O$.

ADVANCED STUDY ASSIGNMENT

Hand-draw a spectrophotometer. Label the components and explain briefly operation of the instrument.

DATA COLLECTION SHEET

DATA COLLECTION SHEET

DATA COLLECTION SHEET

DATA COLLECTION SHEET

DATA COLLECTION SHEET

PART 4

EXPERIMENTS FOR
HAZARDOUS WASTE

11

DETERMINATION OF THE COMPOSITION OF UNLEADED GASOLINE USING GAS CHROMATOGRAPHY

Purpose: To learn to use a capillary column gas chromatography system

To learn to use column retention times to identify compounds

To learn to calibrate a gas chromatograph and quantify the mass of each peak

BACKGROUND

Petroleum hydrocarbons may well be the most ubiquitous organic pollutant in the global environment. Every country uses some form of hydrocarbons as a fuel source, and accidental releases result in the spread and accumulation of these compounds in water, soil, sediments, and biota. The release of these compounds from underground storage tanks is the most common release to soil systems, and this is discussed in Chapter 16. The drilling, shipping, refining, and use of petroleum products all account for serious releases to the environment.

Crude oil consists of straight-chained and branched aliphatic and aromatic hydrocarbons. Upon release into the environment, some compounds undergo oxidation. Chemical and photochemical oxidation occur in the atmosphere; in water and soil systems, microorganisms are responsible for the oxidation. The analysis of crude oil, and organic compounds in general, has improved enormously with the advent of capillary column gas chromatography. In fact, capillary

Environmental Laboratory Exercises for Instrumental Analysis and Environmental Chemistry
By Frank M. Dunnivant
ISBN 0-471-48856-9   Copyright © 2004 John Wiley & Sons, Inc.

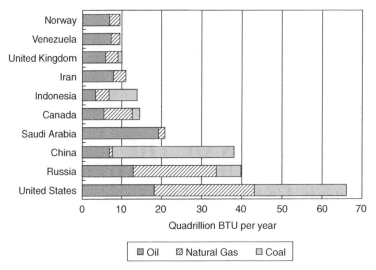

Figure 11-1. Energy production of selected countries. (U.S. EPA, 2002.)

column GC can even identify the country of origin of a crude oil sample based on the chemical/compound composition.

One of the largest problems with respect to the release of hydrocarbons in the environment is that they are hydrophobic (they do not like to be in water). Hydrocarbons are organic compounds and do not undergo hydrogen bonding, and thus do not readily interact with water. As a result, hydrocarbons bioaccumulate in the fatty tissue of plants and animals or associate with organic matter in soils and sediments. Compounds can be toxic at low levels, one of the most common examples being benzene, present in all gasoline products.

Our use of petroleum hydrocarbons is ever-increasing. Figure 11-1 summarizes the production rates for the highest-energy-consuming countries. You will note that the United States produces (and consumes) the most energy per year. But how do we use this energy? Figure 11-2 shows a breakdown of the energy use into

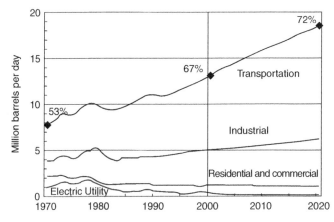

Figure 11-2. Current and predicted energy consumption in the United States. (U.S. EPA, 2002.)

electric, residential and commercial, industrial, and transportation. Transportation, the largest form of consumption, is increasing at an alarming rate. This not only explains the intensive research programs in fuel cell technology but also the geopolitical conflicts in the Middle East.

THEORY

Although it takes months to years to become a good chromatographer, this laboratory exercise will introduce you to the basics of chromatography. There are many highly technical parts to a capillary column GC, including the ultrapure carrier and makeup gases, flow controller values, injector, column, oven, a variety of detectors, and a variety of data control systems. You should consult a textbook on instrumental methods of analysis for details on each of these systems. The basic theory important to understand for this laboratory exercise is that there is generally a separation column for every semivolatile compound in existence. We limit the GC technique to volatile or semivolatile compounds since the compound must travel through the system as a gas. Nonvolatile or heat-sensitive compounds are normally analyzed by high-performance liquid chromatography (HPLC). Compounds are separated in the GC (or HPLC) column by interacting (temporarily adsorbing) with the stationary phase (the coating on the inside wall of the column). The more interaction a compound undergoes with the stationary phase, the later the compound will elute from the column and be detected. This approach allows for the separation of both very similar and vastly different compounds. Vastly different compounds can be separated by relying on the diversity of intermolecular forces available in column coatings (hydrogen bonding, dipole interactions, induced dipole interactions, etc.). Similar compounds are separated using long columns (up to 60 m).

The most important parameter we have for separating compounds in GC is the oven temperature program. If we analyze a complex mixture of compounds at a high temperature (above the boiling point of all of the compounds in the mixture), we do not get adequate separation, and the mixture of compounds will probably exit the system as a single peak. But if we take the same mixture and start the separation (GC run) at a low temperature and slowly increase the oven temperature, we will usually achieve adequate separation of most or all of the compounds. This works by gradually reaching the boiling point (or vaporization point) of each compound and allowing it to pass through the column individually. In this manner, very similar compounds can be separated and analyzed.

You will be using external standard calibration for your analysis. This is the common way that standards are analyzed, in which you analyze each concentration of standard separately and create a calibration curve using peak height or peak area versus known analyte concentration. However, capillary column GC requires that you account for errors in your injections. This is accomplished by having an internal standard, in our case decane, at the same concentration in every sample and standard that you inject. By having the same concentration in every injection, you can correct for injection losses. (The peak area for the decane sample should be the same; if it is not, modern GC systems correct for any losses.)

For a good summary of the theory and use of a gas chromatography system, see the down loadable GC Tutorial (`http://www.edusoln.com`). Your instructor will have this available on a computer for your viewing.

REFERENCE

U.S. EPA, `http://www.epa.org`, accessed July 2003.

IN THE LABORATORY

This laboratory is divided into two exercises. During the first laboratory period, you will determine the retention times of analytes in an unleaded gasoline sample. For the second laboratory period, you will measure the concentration of several components in the gasoline using external and internal standard calibration.

Safety and Precautions

- Safety glasses should be worn at all times during the laboratory exercise.
- This laboratory uses chemicals that you are exposed to every time you fill your car with gasoline. But this does not reduce the toxic nature of the compounds you will be handling. Many of these are known carcinogens and should be treated with care.
- Use all chemicals in the fume hood and avoid inhaling their vapors.
- Use gloves when handling organic compounds.

Chemicals and Solutions

- One or more unleaded gasoline samples
- Neat samples of *m*-xylene, *o*-xylene, benzene, ethyl benzene, isooctane, toluene, and *n*-heptane

Equipment and Glassware

- Several class A volumetric flasks
- 10-, 50-, 100-, and 500-μL syringes for making dilutions
- 1-, 5-, and 10-mL pipets
- a column gas chromatograph equipped with a DB-1 or HP-1 capillary column (a DB-5 or HP-5 will also work, but retention times will change

GC Settings

- Splitless for the first 2 minutes, split mode for the remainder of the run
- Injector temp.:   250°C
- Detector temp.:   310°C
- Oven: Initial temp.:   40°C
                  Hold for: 5 minutes
                  Ramp: 10 to 200°C
                  Hold for: 20 minutes or less

PROCEDURE

Week 1: Determining the Retention Times

1. Turn on the GC, adjust all settings, and allow the instrument to go through a blank temperature run to clean the system. You may also inject pure pentane for this run.

2. While the GC completes the first blank run, prepare a set of reference standards for determining the retention times on your instrument (with the temperature program given in the equipment and glassware section). You will be using decane (C-10) as your internal standard for all solutions. Absolute retention times may vary slightly between GC runs, and the internal standard will allow you to calculate relative retention times (relative to that of decane) and allow you to identify each peak in subsequent GC runs. This first set of standards does not have to be quantitative since you are only checking the retention time, not the concentration of compound in any of the mixtures. To make the standards, place 2 drops of each compound in an individual vial, and add 2 drops of decane and 5 to 10 mL of pentane to each vial. Pentane serves as a good dilution solvent for this procedure since it is very volatile and will exit the GC early to leave a clean window for your analytes to elute.

3. Analyze each solution using the same temperature program and determine the absolute retention time and the relative retention time with respect to decane.

4. Copy the chromatographs for each member in your group and place them in your laboratory manual.

5. There will be plenty of time to spare during this laboratory period, but in order to finish on time, you should keep the GC in use constantly. While you are waiting for each GC run to finish, you should make your quantitative standards for next week's lab. If you wait until next week to make these standards, you will be leaving lab very late. These standards will contain all of your compounds in each solution, but at different concentrations. Analyte concentrations should be 2, 5, 10, 15, and 25 mg/L in pentane. Each solution must also contain the internal standard, decane (at 30 to 50 mg/L). The internal standard will allow you to identify each analyte based on relative retention time and allow you to correct for any injection errors (see the theory section). Seal the standards well and store them in the refrigerator.

Week 2: Determining the Composition of Unleaded Gasoline

1. Turn on the GC, adjust all settings, and allow the instrument to go through a blank temperature run to clean the system. You may also inject pure pentane for this run.

2. While the GC completes the first blank run, arrange a set of reference standards for determining the retention times on your instrument (with the temperature program given in the equipment and glassware section). Since you used pentane as your solvent, some may have evaporated. Allow your standards to come to room temperature and adjust the volume of pentane in each vial. It is unlikely that any of the other compounds evaporated since pentane is the most volatile compound in the mixture, so you do not have to worry about a change in the concentration of your analytes.

3. Make a pure pentane injection, followed by each standard. Run the standards from low to high concentration. Calibrate the GC or store the chromatograms and use your linear least squares spreadsheet.

4. While the standards are running, make dilutions of the pure gasoline for analysis on the GC. Prepare 100- and 250-mg/L solutions of your gasoline in pentane. You will need only a few microliters of this solution, so do not waste solvent by preparing large volumes.

5. Determine the concentration of each analyte in your samples.

6. While you are waiting for the GC runs to finish, your instructor may have some literature work for you. If not, enjoy the free time and clean the lab.

Waste Disposal

Dispose of all wastes in an organic solvent waste container.

ASSIGNMENT

1. Prepare a labeled chromatogram of a midrange calibration standard.
2. Summarize the concentrations of analytes in your gasoline sample and correct for the internal standard.

ADVANCED STUDY ASSIGNMENT

1. Research the operation of a gas chromatograph in the library or on the Internet. Draw and explain each major component of a capillary column system.

2. How does temperature programming affect the elution of compounds from the GC system?

DATA COLLECTION SHEET

DATA COLLECTION SHEET

DATA COLLECTION SHEET

DATA COLLECTION SHEET

DATA COLLECTION SHEET

12

PRECIPITATION OF METALS FROM HAZARDOUS WASTE

Erin Finn

Purpose: To treat a diluted electroplating bath solution for copper, nickel, or chromium using a variety of methods

To learn to use a flame atomic absorption spectrometer

BACKGROUND

Hazardous waste is defined as waste containing one of 39 chemicals specified as hazardous due to their toxic, carcinogenic, mutagenic, or teratogenic properties. The U.S. Environmental Protection Agency (EPA) estimates that 6 billion tons of hazardous waste is created in the United States each year, but only 6% of that, some 360 million tons, is regulated. The remainder is composed of unregulated military, radioactive, small generator (<220 lb per month), incinerator, and household waste. The United States is the largest gross and per capita producer of hazardous waste in the world. Electroplating and engraving operations are one source of this waste. Electroplating baths are used to deposit a thin layer of metal a few millimeters thick onto a metal substrate. These layers may be used to alter the physical properties of a metal surface, such as corrosion resistance, ductile properties, and hardness, or for decorative purposes. The quality of the deposit is affected by the temperature, current, and pH of deposition, as well as the concentration of metal in the bath.

Environmental Laboratory Exercises for Instrumental Analysis and Environmental Chemistry
By Frank M. Dunnivant
ISBN 0-471-48856-9   Copyright © 2004 John Wiley & Sons, Inc.

The most commonly used nickel-plating bath is the Watts bath, which you will use in this experiment. Nickel and chromium plating are often used in conjunction, although the two baths are not mixed, due to the resulting decrease in the quality of the chromium deposits. As metal is deposited over time, the concentration of metal in the bath is decreased to below the optimal concentration, and the bath becomes less effective. It is at this time that the bath must be disposed of or regenerated, and it is the disposal process with which we are concerned. A common initial step in the treatment of such wastes is dilution by emptying the vat into a large pool of water. In this case, the electroplating solutions are diluted to 1 : 50 from average starting plating bath concentrations because this is the greatest dilution that can readily be achieved without having to make large excesses of solution or perform serial dilutions.

Various methods of treatment exist, depending on the composition and concentration of the solution to be treated. One of the cheapest and most universal treatment methods is pH precipitation, which you will perform on nickel and copper. Precipitation by pH works on the principle that at high pH values, metals form their insoluble hydroxides; for example,

$$Cu^{2+} + 2\,OH^- \rightarrow Cu(OH)_2(s)$$
$$Ni^{2+} + 2\,OH^- \rightarrow Ni(OH)_2(s)$$

Unfortunately, this method has a disadvantage: Each metal has a unique pH value at which its hydroxide is least soluble and therefore most effectively precipitated. Literature values are presented in Table 12-1. At pH values above this ideal pH, the solubility actually increases again as the metal coordinates to form charged hydroxide species. This makes pH precipitation of mixed metal solutions difficult. Additionally, although it can be effective, pH precipitation is not always as easy to regulate consistently as are other methods. This method is also effective in treating chromium and is therefore not used in this experiment to treat hexavalent chromium. The value presented in Table 12-1 is for chromium(III), and pH precipitation would first require reduction of the chromium and then adjustment of the pH.

Another method of water treatment is the use of ferric chloride ($FeCl_3$). This operates by a completely different mechanism known as *coagulation*. Coagulation is a method to improve settling rates by increasing the size and specific gravity of

TABLE 12-1. Literature Values of Optimum pH for Precipitation of Metal Ions

Metal	Optimum pH
Cr(III)	7.5
Cu	8.1[a]
Ni	10.8
Mixed metals given above	8.5

[a]Although this is the ideal literature value, it has been found in designing this exercise that 8.6 is a more effective pH value for precipitation of copper.

a particle. It can be used to remove silt, clays, bacteria, minerals, and oxidized metals and other inorganics from waters. The Fe^{3+} in ferric chloride reacts with hydroxide in basic solution:

$$Fe^{3+} + 3\,OH^- \rightarrow Fe(OH)_3(s)$$

Iron(III) hydroxide forms a colloid-sized particle (0.001 to 1 μm) that complexes with water molecules and becomes negatively charged by coordination of the iron with anions, especially hydroxide, in solution. Positively charged metal ions bind multiple negatively charged colloidal particles together and create a large body that precipitates out of solution and can easily be separated via sand filtration, or if sufficient time is available, even settling. Either of these methods is effective in generating a clear supernatant layer from the coagulated solution; sand and gravel filtration are common techniques used to treat water and effluent because filtration is cheap and requires fairly low maintenance. Ferric chloride is a convenient coagulant because it is cheap, easy to use, and works well over a wide pH range. It is important that the pH be high enough to counteract the acidic nature of electroplating baths and the acidity of the iron in solution, which acts as a Lewis acid to cause water to dissociate. This treatment was not found to be effective with hexavalent chromium, however.

An effective treatment of hexavalent chromium involves ferrous chloride, which accomplishes reduction and precipitation simultaneously in nearly neutral to slightly basic solutions. Note that the pH given in Table 12-1 for Cr^{3+} is within the neutral range required. The reduction reaction is

$$4\,H_2O + CrO_4^{2-} + 3\,Fe^{2+} + 4\,OH^- \rightarrow 3\,Fe(OH)_3(s) + Cr(OH)_3(s)$$

A mixed iron–chromium solid in the form $Fe_xCr_{1-x}(OH)_3$ is also reported to be formed, where x is 0.75 when the stoichiometric relationship described above is applied.

$$4\,H_2O + CrO_4^{2-} + 3\,Fe^{3+} + 4\,OH^- \rightarrow 4\,Fe_{0.75}Cr_{0.25}(OH)_3(s)$$

This treatment, in combination with ferric chloride treatment, can be used to process a solution of mixed metal waste containing copper, nickel, and chromium. Although in actual practice chromium is not often mixed with other metals due to the detrimental effect that this has on chromium bath efficiency, all of these metals could be present in a hazardous waste treatment situation.

THEORY

The driving mechanism behind the effectiveness of precipitation treatments is the solubility product. You may recall from general chemistry that the *solubility product* is defined as the product of the concentrations of the ions involved in an equilibrium, each raised to the power of its coefficient in the equilibrium equation.

The equilibrium referred to is that between a saturated solution of a compound and the solid form of that compound. Compounds with a low solubility product do not dissolve to any great extent in water, and may be considered insoluble. Compounds with a high solubility product, such as potassium perchlorate, dissolve readily in water. The solubility product for potassium perchlorate can be expressed as

$$k_{spKClO_4} = [K^+][ClO_4^-] = 1.05 \times 10^{-2}$$

The solubility product of lead(II) chloride is

$$k_{spPbCl_2} = [Pb^{2+}][Cl^-]^2 = 1.70 \times 10^{-5}$$

while the solubility product of lead(II) hydroxide is

$$k_{spPb(OH)_2} = [Pb^{2+}][OH^-]^2 = 1.43 \times 10^{-20}$$

The difference in k_{sp} between lead(II) chloride and lead(II) hydroxide illustrates the reason that precipitation by pH is effective at removing metals from solution.

REFERENCES

Brown, T. L., H. E. Lemay, B. E. Bursten, and J. R. Burfge, *Chemistry: The Central Science,* 8th ed., Prentice Hall, Upper Saddle River, NJ, 2000, p. 660.

Guidance Manual for Electroplating and Metal Finishing Pretreatment Standards, U.S. EPA, Feb. 1984, http://www.epa.gov/npdes/pubs/owm0022.pdf, accessed Feb. 2003.

Hazardous Waste, http://www.members.tripod.com/recalde/lec6.html, accessed May 2003.

http://www.waterspecialists.biz/html/precipitation_by_pH_, accessed Feb. 2003.

Lide, D. R. and H.P.R. Frederikse (eds.), *CRC Handbook of Chemistry and Physics*, CRC Press, Boca Raton, FL, 1997, pp. 8-106 to 8-109.

WTA's World Wide Water, "Coagulation," http://www.geocities.com/capecanaveral/3000/coag.htm, accessed May 2003.

IN THE LABORATORY

The overall goal of all of these treatments is to remove as much of the metal as possible. In industry your target removal level would be the maximum emission concentration allowed by a state or federal governing body. The EPA has established Pretreatment Standards for Existing Sources (PSES) of electroplating waste in the *Guidance Manual for Electroplating and Metal Finishing Pretreatment Standards*, based on the requirements of subchapter N of the *Code of Federal Regulations*, Chapter 1. These standards limit the concentration of hazardous waste components that may be present in the wastewater effluent of electroplating operations. For a facility discharging >38,000 L/day, the limits are as shown in Table 12-2. These limits were established in 1984 and are part of the National Pollutant Discharge Elimination System (NPDES) limits that regulate effluents. For facilities discharging <38,000 L/day, none of these metals are regulated.

Safety Precautions

- Keep in mind that while the plating baths are diluted about 50-fold, they are still considered hazardous waste (notice the colors—brightly colored solutions are usually not a good sign unless they are indicators!). This means that they must *not* be dumped down the drain without treatment!
- The copper-plating bath especially is quite acidic (pH about 1.5), as you will notice when you pH-treat it. Be careful not to spill on yourself!
- Keep a waste beaker for all your plating bath waste. When you are finished, estimate its volume and try to treat any remaining waste.
- All precipitates should be collected in waste jars.
- Supernatants and filtrates should be clean enough to meet EPA standards by the time you are finished, and can then be dumped down the drain with excess water. Be sure that you check the pH and confirm that they meet standards by checking them first on the AAS unit.

Chemicals and Solutions

Each student or group will be assigned one metal to work with. The solutions required for each group are slightly different.

TABLE 12-2. EPA Pretreatment Standards for Existing Sources

Metal	Daily Maximum (mg/L)	Max. 4-Day Average (mg/L)
Total Cr	7.0	4.0
Total Cu	4.5	2.7
Total Ni	4.1	2.6

Group 1: Copper

- 100 mL of copper-plating bath: 1.5 g of $CuSO_4 \cdot 5H_2O$
  5.6 mL of concentrated H_2SO_4
  Deionized water
- 25 mL of 1.3 M ferric chloride
- 200 mL of 2 M sodium hydroxide
- 1% Nitric acid for preparing samples for FAAS
- Glass wool
- A few grams of sand

Group 2: Nickel

- 100 mL of nickel-plating bath: 22.8 g of $NiSO_4 \cdot 6H_2O$
  6.8 g of $NiCl_2 \cdot 6H_2O$
  3.7 g of H_3BO_3
  Deionized water
- 25 mL of 1.3 M ferric chloride
- 50 mL of 2 M sodium hydroxide
- 1% Nitric acid for preparing samples for FAAS
- Glass wool
- A few grams of sand

Group 3: Chromium

- 100 mL of chromium-plating bath: 0.3 g of CrO_3
  0.003 g of Na_2SO_4
  Deionized water

(*Note*: A serial dilution is required to get the correct quantity of sodium sulfate, because you cannot weigh out 3 mg accurately.)

- 25 mL of 1 M ferrous chloride
- 100 mL of 2 M sodium hydroxide
- 10.00 mL of nickel bath and 10.00 mL of copper bath, to be obtained from the other groups
- 25 mL of 1.3 M ferric chloride
- 1% Nitric acid for preparing samples for FAAS
- Glass wool
- A few grams of sand

Equipment and Glassware

- 10-, 25-, 50-, and 100-mL volumetric flasks
- Graduated cylinders
- Pipets
- Glass chromatography columns (20 mm or wider) with buret clamps and ring stands
- Beakers
- 50- and 125-mL Erlenmeyer flasks
- Long glass stir rods
- Scintillation vials (four per person or group)
- Stir plates and beans
- pH meter and buffer solutions
- FAAS with Ni, Cu, and Cr hollow cathode lamps

PROCEDURE

Group 1: Copper

You will treat your waste by pH precipitation and by ferric chloride coagulation. First, make your solutions as described above. You will want to start making the copper solution early because it takes some time to dissolve. The ferric chloride also takes a little while but dissolves within 5 minutes on a stir plate. It does, however, foam on top, preventing a good volume reading. Simply do your best to get the volume as close as possible to the desired total. Since you will be dispensing the ferric chloride solution with a graduated cylinder—it is too thick and foamy to use a pipette and could cause clogging—the error introduced in doing this is one of many.

pH Precipitation. Pipet 25.00 mL of your copper bath into an Erlenmeyer flask. Adjust the pH to 8.6 using 2 M NaOH. This adjustment can be difficult, as the pH changes are very sensitive near the neutral range. You may wish to dilute your sodium hydroxide to make the changes easier to fine tune. Using 2 M NaOH, it should take about 40 to 45 mL. Since the copper solution already contains sulfuric acid, 1 or 2 drops of very dilute sulfuric acid (about 0.1 M) may be used to correct the pH if you overshoot a pH of 8.6. Cover the treated solution and allow it to settle until next week's lab. If you desire to continue working now, wait a few minutes and it will settle, but be sure that the supernatant is *clear* before proceeding. Pipet off a few milliliters of supernatant, being careful not to disturb the precipitate. For FAAS analysis, mix 3.00 mL of supernatant with 3.00 mL of 1% HNO_3.

FeCl₃ Treatment. Pipet 25.00 mL of copper solution into a flask. Add approximately 5 mL of 1.3 M FeCl₃ and 45 mL of 2 M NaOH. In both cases, it is better to err on the side of adding too much rather than too little. However, if you add excess FeCl₃, be sure to compensate for it with excess NaOH. It is imperative that the solution be basic for the treatment to work. You may wish to confirm this using litmus paper or universal indicator paper. You may stop here with your solution covered until the next lab period if desired, or continue working.

The next step is to construct a sand column. Use a glass rod to push a small plug of glass wool to the bottom of the column. Then add about 2 cm of sand over the top. Tap and gently shake the column to allow the sand to settle and reduce air gaps. Smoothly pour your treated solution onto the column. It is helpful to try to pour just the liquid initially, so that the initial stages of filtration will proceed more quickly. Once the solid plugs the pores in the sand, filtration takes much longer; it may take a couple of hours for the supernatant to filter through completely. Collect the filtrate in a clean beaker. For FAAS analysis, pipette 3.00 mL of supernatant and 3.00 mL of 1% HNO_3 into a scintillation vial.

During the second week of lab, you will analyze your samples for copper using FAAS. You will need to begin by making calibration standards at 2, 4, 8, 20, and 40 ppm (this range may depend on the FAAS unit you use) in copper with the corresponding quantities of sulfuric acid. You will probably need to use serial dilutions. Remember to make your standards in 1% nitric acid instead of deionized water. When ready to do your analyses, warm up the instrument as instructed and create your calibration curve. Use 1% nitric acid as your blank. You will share this calibration curve with the students who are working with the mixed-chromium wastewater; they will need it to analyze their mixed waste treatment. Analyze your samples five times each. You should also try to coordinate timing so that the chromium students can analyze their treated mixed waste while the correct lamp is installed in the instrument and is warmed up.

Group 2: Nickel

You will treat your waste by pH precipitation and by ferric chloride coagulation. First, make your solutions as described earlier. You will want to start making the nickel solution early because it takes some time to dissolve. The ferric chloride also takes a little while but dissolves within 5 minutes on a stir plate. It does, however, foam on top, preventing getting a good volume reading. Simply do your best to get the volume as close as possible to the desired total. Since you will be dispensing the ferric chloride solution with a graduated cylinder—it is too thick and foamy to use a pipette and could cause clogging—the error introduced in doing this is one of many.

pH Precipitation. Pipet 25.00 mL of your nickel bath into an Erlenmeyer flask. Adjust the pH to 10.8 using 2 M NaOH. It should take approximately 5 to 7 mL. Since the nickel solution already contains nickel(II) sulfate, 1 or 2 drops of dilute sulfuric acid (<1 M) may be used to correct the pH if you overshoot the pH value of 10.8. Cover the treated solution and allow it to settle until next week's lab. If you desire to continue working now, wait a few minutes and it will settle, but be sure the supernatant is *clear* before proceeding. Then pipet off a few milliliters of supernatant, being careful not to disturb the precipitate. For FAAS analysis, mix 3.00 mL of supernatant with 3.00 mL of 1% HNO_3.

FeCl₃ Treatment. Pipet 25.00 mL of nickel solution into a flask. Add 7 mL of 1.3 M $FeCl_3$ and 20 mL of 2 M NaOH. In both cases it is better to err on the side of adding too much rather than too little. However, if you add excess $FeCl_3$, be sure to compensate for it with excess NaOH. It is imperative that the solution be basic for the treatment to work. You may wish to confirm the basicity of the solution using litmus paper or universal indicator paper. You may stop here with your solution covered until the next lab period if desired, or continue working.

The next step is to construct a sand column. Use a glass rod to push a small plug of glass wool to the bottom of the column. Then add about 2 cm of sand over the top. Tap and gently shake the column to allow the sand to settle and reduce air gaps. Smoothly pour your treated solution onto the column. It is helpful to try to pour just the liquid initially, so that it can pass through more quickly. Once the solid blocks the pores in the sand, filtration takes much longer; it may take a couple of hours for the supernatant to finish coming through the sand. Collect the filtrate in a clean beaker. For FAAS analysis, pipet 3.00 mL of supernatant and 3.00 mL of 1% HNO_3 into a scintillation vial.

During the second week of lab, you will analyze your samples for nickel using FAAS. You will need to begin by making calibration standards at 2, 4, 8, 20, and 40 ppm (this range may depend on the FAAS unit you use) in total nickel, with $NiSO_4$ and $NiCl_2$ composing appropriate proportions of the total. These standards should contain correspondingly appropriate quantities of boric acid so that the matrix is the same for your standards as the matrix of your waste solution. You will probably need to use serial dilutions. Remember to make your standards in 1% nitric acid instead of deionized water. When ready to do your analyses, warm up the instrument as instructed and create your calibration curve. Use 1% nitric acid as your blank. You will share your calibration curve with the students who are working with the mixed chromium wastewater; they will need it to analyze their mixed waste treatment. Analyze your samples five times each. You should also try to coordinate timing so that the chromium students can analyze their treated mixed waste while the correct lamp is installed in the instrument and warmed up.

Group 3: Chromium

You will treat your chromium by ferrous chloride precipitation and will also treat a mixed waste that contains copper and nickel in addition to chromium. First, make your solutions as described above. The ferric chloride takes a little while to dissolve but will do so within 5 minutes on a stir plate. It does, however, foam on top, preventing getting a good volume reading. Simply do your best to get the volume as close as possible to the desired total. Since you will be dispensing the ferric chloride solution with a graduated cylinder—it is too thick and foamy to use a pipette and could cause clogging—the error introduced in doing this is one of many.

FeCl₂ Precipitation. Pipet 25.00 mL of chromium solution into a flask. Add 5 mL of 1 *M* $FeCl_2$ and 5 mL of 2 *M* NaOH. In both cases it is better to err on the side of adding too much rather than too little. However, if you add excess $FeCl_2$, be sure to compensate for it with excess NaOH. For the treatment to work, it is imperative that the solution be basic. You may wish to confirm the basicity of the solution using litmus paper or universal indicator paper. You may stop here with your solution covered until the next lab period if desired, or continue working.

The next step is to construct a sand column. Use a glass rod to push a small plug of glass wool to the bottom of the column. Then add 2 cm of sand over the top. Tap and gently shake the column to settle the sand and reduce air gaps.

Smoothly pour your treated solution onto the column. It is helpful to try to pour just the liquid initially, so that the initial stages of filtration proceed more quickly. Once the solid fills the pores in the sand, filtration takes much longer; it may take a couple of hours for all of the supernatant to come through. Collect the filtrate in a clean beaker. For FAAS analysis, pipet 3.00 mL of filtrate and 3.00 mL of 1% HNO_3 into a scintillation vial.

Mixed Waste Treatment. Prepare a mixed electroplating bath waste by pipetting 10.00 mL of each metal solution into a flask. You will need to get copper and nickel bath solutions from the other groups. Add 5.5 mL of 1.3 M $FeCl_3$ and 30 mL of 2 M sodium hydroxide. Mix the solution well and allow it to sit. You may stop here or after the filtration step that follows. While it sits, construct a sand column as you did before, with glass wool and 2 cm of sand in the bottom. Pour your treated solution slowly over the top of the column. Collect the filtrate. You will notice that it is a bright yellow color. This is because the ferric chloride has succeeded in removing the nickel and copper but not the chromium. To remove the chromium, you will need to add 2.2 mL of 1 M ferrous chloride and 2.2 mL of 2 M sodium hydroxide and swirl to mix well. Once this is done, you may stop here or continue. Allow the precipitate to settle and collect a few milliliters of supernatant carefully with a pipette so as to avoid disturbing the precipitate. Mix 3.00 mL of supernatant with 3.00 mL of 1% HNO_3 for FAAS analysis.

During the second week of lab, you will analyze your ferrous chloride–treated sample for chromium and your mixed waste–treated sample for copper, nickel, and chromium using FAAS. For the mixed waste, it does not matter in what order you analyze for the various metals. You will need to coordinate instrument time with other students so as to be able to perform your analyses while the appropriate lamp is installed and warmed up in the instrument. You will need to begin by making calibration standards at 1.6, 4, 8, 20, and 40 ppm in chromium (this range may depend on the FAAS unit you are using). You will need to use serial dilutions. For the copper and nickel analyses, you will use the calibration curves created by your peers. Remember to make your standards in 1% nitric acid instead of deionized water. When ready to do your analyses, warm up the instrument as instructed and create your calibration curve. Use 1% nitric acid as your blank. Analyze each sample five times.

You should carefully plan your data collection and recording strategy since there are several types of data to be collected and the entire class is dependent on your data. After collecting your FAAS results, you should perform a linear least squares analysis, convert absorbance signal to concentration, and then correct that concentration for the dilution you used in preparing your supernatant sample for FAAS, to determine the concentration of metal in your treated solution. Then correct for dilutions during treatment (assuming additive volumes) and calculate your percent removal. (Contrast your results for pH precipitation in light of the calculated solubility of the metals based on the final solution pH and the K_{sp} value of the hydroxide of that metal. Why might the two answers not agree?)

Questions to think about for your write-up:

1. How effective is each treatment for each metal? Do the treated solutions meet EPA standards?
2. How reproducible are the results of the treatment when the same procedure is used?
3. Which procedure is easiest to use on this scale? On an industrial scale (i.e., treating at least 100 L of effluent)?
4. Which procedure is cheapest? Which uses the least harmful chemicals?

PRECIPITATION OF METALS FROM HAZARDOUS WASTE: DATA COLLECTION SHEET 1

Name: _____

Lab partners: _____

Date: _____

Metal assigned: _____

Solution Preparation

Metal solution: Cu_____   Metal solution: Ni_____

Volume prepared: 100 mL_____   Volume prepared: 100 mL_____

Solution Component	Mass or Volume Used

Solution Component	Mass or Volume Used

Metal solution: Cr_____   Mixed metals solution

Volume prepared: 100 mL_____   Volume prepared: 30.00 mL_____

Solution Component	Mass or Volume Used

Solution Component	Mass or Volume Used

Solution	Mass of Solid or Volume Used	Concentration
NaOH		
$FeCl_3 \cdot 6H_2O$		
$FeCl_2$		
HNO_3		

Data Collection

pH Precipitation of Nickel

Trial	Initial pH	Initial Buret Vol. (mL)	Final Buret Vol. (mL)	Vol. of 2 M NaOH Used (mL)	Final pH
1					
2					
3					

FeCl₃ Coagulation of Nickel

Trial	Vol. of FeCl$_3$ Used (mL)	Vol. of 2 M NaOH Used (mL)
1		
2		
3		

PRECIPITATION OF METALS FROM HAZARDOUS WASTE: DATA COLLECTION SHEET 2

FeCl₂ Precipitation of Cr

Trial	Vol. of $FeCl_2$ Used (mL)	Vol. of 2 M NaOH Used (mL)
1		
2		
3		

Mixed Waste Treatment

Trial	Vol. $FeCl_3$ (mL)	Vol. NaOH (mL)	Vol. $FeCl_2$ (mL)	Vol. NaOH (mL)
1				
2				
3				

FAAS Data

Element: _____

Wavelength: _____

Slit width: _____

Lamp current: _____

Fuel flow: _____

Oxidant flow: _____

Sample	Instrument Signal (absorbance)						Conc. (ppm)	Corr. Conc.
	1	2	3	4	5	Avg.		
Blank (1% HNO$_3$)								
2 ppm								
4 ppm								
8 ppm								
20 ppm								
40 ppm								
pH precipitated (1)								
pH precipitated (2)								
pH precipitated (3)								
Coagulated (1)								
Coagulated (2)								
Coagulated (3)								
Mixed waste (1)								
Mixed waste (2)								
Mixed waste (3)								

Linear Least Squares Results

r^2: _____

m: _____

b: _____

PRECIPITATION OF METALS FROM HAZARDOUS WASTES: DATA COLLECTION SHEET 3

Element: _____   Lamp current: _____

Wavelength: _____   Fuel flow: _____

Slit width: _____   Oxidant flow: _____

Sample	Instrument Signal (absorbance)						Conc. (ppm)	Corr. Conc.
	1	2	3	4	5	Avg.		
Blank (1% HNO_3)								
1.6 ppm								
4 ppm								
8 ppm								
20 ppm								
40 ppm								
$FeCl_2$ (1)								
$FeCl_2$ (2)								
$FeCl_2$ (3)								
Mixed waste (1)								
Mixed waste (2)								
Mixed waste (3)								

Linear Least Squares Results

r^2: _____

m: _____

b: _____

PRECIPITATION OF METALS FROM HAZARDOUS WASTE: DATA COLLECTION SHEET 4

Element: _____ Lamp current: _____

Wavelength: _____ Fuel flow: _____

Slit height: _____ Oxidant flow: _____

Sample	Instrument Signal (absorbance)						Conc. (ppm)	Corr. Conc.
	1	2	3	4	5	Avg.		
Blank (1% HNO$_3$)								
2 ppm								
4 ppm								
8 ppm								
20 ppm								
40 ppm								
pH precipitated (1) (pH 8.64)								
pH precipitated (2)								
pH precipitated (3)								
Coagulated (1)								
Coagulated (2)								
Coagulated (3)								
Mixed waste (1)								
Mixed waste (2)								
Mixed waste (3)								

Linear Least Squares Results

r^2: _____

m: _____

b: _____

Optional Unknown

Treatment method: _____   Vol. of unknown treated: _____

Dilution factor for FAAS: _____   Vol. of treatment solution used: _____

Sample	Instrument Signal (absorbance)						Conc. (ppm)	Corr. Conc.
	1	2	3	4	5	Avg.		
Dil. unknown								
Treated 1								
Treated 2								
Treated 3								

DATA COLLECTION SHEET

DATA COLLECTION SHEET

DATA COLLECTION SHEET

DATA COLLECTION SHEET

DATA COLLECTION SHEET

13

DETERMINATION OF THE NITROAROMATICS IN SYNTHETIC WASTEWATER FROM A MUNITIONS PLANT

Purpose: To determine the concentration of nitroaromatic compounds in munitions wastewater

To learn to use a high-performance liquid chromatograph

BACKGROUND

Abandoned ammunition plants from World War II litter the United States and Europe, as well as many other countries. Waste from these plants primarily contaminates the soil, but leachate is released during rain and snowmelt events. Examples of these sites in the United States include the Iowa Army Ammunitions Plant (Middleton, Iowa), Fort Hill (Washington, DC), and the Red Stone Arsenal (Huntsville, Alabama). The primary compounds in the leachate from these sites, designated as hazardous waste by most countries, are trinitrotoluene (TNT), cyclotrimethylene–trinitramine (RDS), cyclotetramethyulene–tetratrinitramine (HMS), and a variety of nitro-substituted benzenes and toluenes. TNT is photo-active, producing a pink color in surface wastewaters, and is commonly referred to as *pink water* (our solutions will be yellow, due to the compounds we use). The total concentration of nitroaromatic compounds in these waste streams can reach several hundred parts per million. These wastewaters are also highly subject to oxidation, producing anilines that are toxic to aquatic organisms.

Environmental Laboratory Exercises for Instrumental Analysis and Environmental Chemistry
By Frank M. Dunnivant
ISBN 0-471-48856-9 Copyright © 2004 John Wiley & Sons, Inc.

THEORY

A variety of techniques are available for measuring the concentration of nitro-aromatics in water. The two most common approaches are gas chromatography (GC) and high-performance liquid chromatography (HPLC). Both of these instruments are ideal for analyzing complex mixtures of analytes. Whereas GC was developed to analyze compounds that were volatile (boiling points less than 300°C) and not subject to thermal degradation in the instrument, high-performance liquid chromatography was developed to analyze nonvolatile compounds and compounds that degraded readily under heat. In many cases, GC and HPLC can be used to analyze the same compounds, as is the case for nitro aromatics. We will be using an HPLC equipped with a UV–Vis detector in this exercise. I refer you to the HPLC tutorial (`http://www.edusolns.com`) for the general operation and theory of this instrument. Since all instruments are slightly different, your instructor will give you a demonstration of the instrument that you will use.

REFERENCE

Agilent Technologies Product Catalog, 2003–2004, `http://www.agilent.com`.

IN THE LABORATORY

This is a relatively straightforward laboratory exercise that illustrates the easy use of the HPLC for water samples. Like the synthetic wastewater sample that your instructor will give you, most waste from contaminated sites is relatively free of matrix effects, with one exception, but proper use of the HPLC requires that your samples be in the same matrix (in our case solvent) as your standards. The gradient (the mobile phase) used in the HPLC is 45% water and 55% methanol. Since your analytical column may not perform exactly as the one used to develop this experiment, I suggest using separate solvent bottles for each solvent. This will allow you to adjust the gradient slightly as needed. Your instructor may have done this beforehand. First, you will make your external standards, containing four nitroaromatic compounds. As you calibrate the HPLC with these external standards, mix your sample with methanol to achieve the same solvent composition as that used in your HPLC gradient. Finally, inject your samples and calculate the concentration of each compound.

Safety Precautions

- Safety glasses must be worn at all times during this laboratory experiment.
- As with any chemical in the laboratory, you should handle these as though they are hazardous. Avoid skin and eye contact and do not breath vapors or chemical dust from the reagents.
- Methanol is flammable and should be used in a fume hood away from flames or hot plates.

Chemicals and Solutions

- HPLC-grade water and methanol
- 1,3-Dinitrobenzene
- Trinitrotoluene (the least expensive source found was Chem Service, West Chester, Pennsylvania)
- 4-Amino-2-nitrotoluene
- 2,6-Dinitrotoluene

Prepare a 5000-mg/L solution of each nitroaromatic compound by dissolving 0.125 g in 25 mL of methanol/water (the composition should match your HPLC gradient).

Equipment and Glassware

- Standard volumetric flasks and pipets
- Isocratic or gradient HPLC with two solvent reservoirs

- Standard C-18 column and precolumn (a 12.5-cm by 4.6-cm column was used to obtain the chromatogram shown in Figure 13-1 and the retention times given in Table 13-1)

TABLE 13-1. Peak Retention Times

Compound	Retention Time (min)
4-Amino-2-nitrotoluene	10.43
1,3-Dinitrobenzene	13.17
Trinitrotoluene	17.49
2,6-Dinitrotoluene	20.12

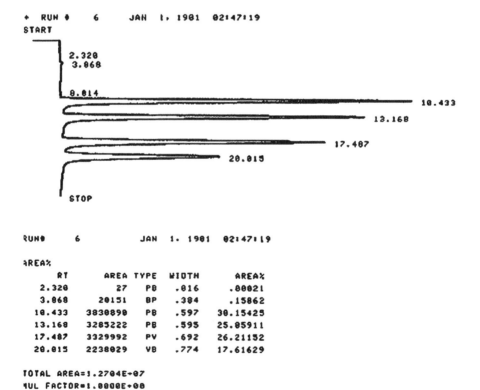

Figure 13-1. Example chromatogram for the analysis of nitroaromatics (about 50 ppm for each compound). (Attenuation setting, 4; chart speed, 0.2 cm/min; flow rate, 0.30 mL/min; 10-μL sampling loop, detection wavelength, 230 nm.)

PROCEDURE

1. Sign in the HPLC logbook, turn on the HPLC, including the UV lamp (set at 230 nm), and allow the instrument to warm up for 5 minutes.
2. Start the gradient (predetermined by your lab instructor) and allow the system to equilibrate while you prepare your standards.
3. Prepare your standards. First, prepare 25 mL of a 5000-mg/L solution (in methanol) of each compound. Next, make 50 to 100 mL of standards containing all compounds. The composition of the solvent should be identical to that of your HPLC gradient. Suggested concentrations are approximately 1, 5, 10, 25, 50, and 100 ppm. You should make your standards accurate to three significant figures.
4. Inject your standards from low to high concentration.
5. Inject a blank (water and methanol) to ensure that the system is not contaminated by your standards.
6. Inject your samples. After you are finished, record any instrument problems in the logbook and sign out.
7. Analyze your data using the linear least squares spreadsheet created Chapter 2 or provided by your instructor.

Waste Disposal

Your samples and waste from the HPLC must be treated as hazardous waste since they contain methanol and nitroaromatic compounds. These should be placed in a glass storage container and disposed of in accordance with federal guidelines.

ASSIGNMENT

Calculate the concentration of each compound in your sample using your linear least squares spreadsheet, accounting for any dilutions you made.

ADVANCED STUDY ASSIGNMENT

1. Draw and label a gradient HPLC system.
2. Describe each major component of the system.

DATA COLLECTION SHEET

DATA COLLECTION SHEET

DATA COLLECTION SHEET

DATA COLLECTION SHEET

DATA COLLECTION SHEET

14

DETERMINATION OF A SURROGATE TOXIC METAL IN A SIMULATED HAZARDOUS WASTE SAMPLE

Purposes: To introduce complex sample matrices

To learn flame atomic absorption spectroscopy techniques for analyzing trace metal solutions

To learn to titrate complex samples using the EDTA titration method

To learn to use solid-state calcium electrodes

To learn to write in a scientific and professional manner

BACKGROUND

The global problem of hazardous waste did not occur overnight. It is documented as early as the Roman Empire with the use of lead 2000 years ago. Early sources of hazardous waste included the smeltering of metal ore and the tanning of animal hides. The industrial revolution brought an onslaught of hazardous waste issues that were not addressed until the 1970s and 1980s. But first, what is *hazardous waste*? Each country has its own definition, but there are remarkable similarities between them. The United Nations Environment Programme, from 1985, summarizes the problem (LaGrega et al., 1994): "Hazardous wastes mean waste [solids, sludges, liquids, and containerized gases] other than radioactive [and infectious] wastes which, by reason of their chemical activity or toxic, explosive,

Environmental Laboratory Exercises for Instrumental Analysis and Environmental Chemistry
By Frank M. Dunnivant
ISBN 0-471-48856-9   Copyright © 2004 John Wiley & Sons, Inc.

corrosive, or other characteristics, cause danger or likely will cause danger to health or the environment, whether alone or when coming into contact with other waste...."

Classic pollutants that are specifically listed as hazardous waste include waste containing DDT, mercury, and PCBs, just to name a few notable chemicals. These and other chemicals have led to highly publicized disasters, such as Love Canal in New York and Times Beach in Missouri. Old abandoned sites such as these fall under the Comprehensive Environmental Response, Compensation, and Liability Act (CERCLA, 1980, and subsequent reauthorizations) commonly known as Superfund, which is designed to clean up abandoned sites. Hazardous wastes being generated today are covered under the Resource Conservation and Recovery Act (RCRA, 1976, and subsequent amendments) that is designed to prevent future disasters such as Time Beach and Love Canal from occurring. Similar programs are in place in the United Kingdom (Poisonous Waste Act of 1972) and in Germany (the solid waste laws of 1976).

In the United States (under RCRA), hazardous wastes are further characterized into the major categories of (1) inorganic aqueous, (2) organic aqueous waste, (3) organic liquids, (4) oils, (5) inorganic sludges and solids, and (6) organic sludges and solids. These categories are very important and determine the final resting place or treatment of waste. For example, some wastes are placed in landfills, but prior to the placement of the waste in a landfill, it must be characterized (i.e., analyzed for the type and quantity (concentration) of toxic compounds). This leads to the focus of this laboratory exercise, the characterization of an inorganic hazardous waste. Actually, due to safety concerns, we will be analyzing a simulated hazardous waste, carbonated beverages. These beverages make excellent simulated hazardous wastes because of their complex matrices (viscosity due to the presence of corn sugar, the presence of phosphates that selectively bind to calcium in the FAAS unit, their color, their pH, and their carbonation). In place of measuring a toxic metal, which you could do easily and safely, we will be analyzing for calcium, both because is present in every carbonated beverage and because it avoids the generation and costly disposal of real hazardous waste.

THEORY

Many chemicals, especially metals, can be analyzed by more than one technique. The focus of this laboratory exercise is to learn the flame atomic absorption spectroscopy (FAAS) unit, but you will also use ethylenediaminetetraacetic acid (EDTA) titration from quantitative analysis and a solid-state calcium electrode. The EDTA titration and solid-state electrode are fairly easy to understand since titrations are standard procedures in chemistry courses and the calcium electrode is only slightly more complicated than the familiar pH electrode. The FAAS unit will need more explanation.

The FAAS unit works basically on the Bohr principle, which explains the light absorbed and emitted from an excited hydrogen atom using the equations

$$E = h\nu$$
$$c = \lambda\nu$$

and Avogadro's number. You used these equations in general chemistry to calculate the wavelength of light emitted by line transitions (see a general chemistry textbook for a review). In heavier elements, there are many more transitions that can occur since there are more electrons and more potential excited energy states. Selectivity or probability rules from physical chemistry allow us to predict which transitions are the most likely, and for most elements there are one to three predominant absorption (for absorption spectroscopy) and emission (for emission spectroscopy) lines. For calcium, the most common absorption line is at 422.7 nm, which yields a detection limit of slightly less than 1 part per million.

The FAAS unit works by first turning your sample into a gaseous cloud containing ground-state gaseous calcium in an acetylene–air flame. Very pure light (for calcium at 422.7 nm) is passed through the gaseous cloud of your sample and the flame. When no calcium is present (in your blank), the light passes through the flame unhindered and no absorption occurs (the light is separated by a wavelength separator and detected by a photomultiplier tube). When calcium is present, the ground-state gaseous atoms absorb some of the 422.7-nm light, and an electron in the calcium is excited to a higher energy level. The absorption of light is related directly to the abundance of atoms in the flame (or concentration in your sample). Thus, you can create a calibration curve of concentration versus absorbance and determine the concentration of calcium in your unknown sample. Of course, there are always complications when you are analyzing samples with a complex matrix. For your sample, you have to be concerned with viscosity effects since your standards are in relatively pure water and your sample is in corn syrup. There may be other elements that interfere with the FAAS, electrode, or titration techniques. Most important, some metals (especially calcium) form inorganic salts with phosphate in the sample that prevent the formation of ground-state gaseous atoms and result in the underestimation of calcium in your analysis.

Two techniques have been developed to address these concerns specifically: standard addition and releasing agents. *Standard addition* is difficult to explain. To begin our discussion, refer to Figure 14-1, which presents the results of a standard addition experiment similar to the one you will be conducting. Remember that the purpose of this approach is to try to minimize the presence of interfering compounds in the sample matrix or to overcome these interferences. We do this by making all of our standards in the sample matrix. First, a set of identical solutions, each containing the same volume of sample, are placed in individual beakers. Increasing masses of standard (calcium) are added to all but one of these solutions. Each sample is analyzed the same way on the FAAS unit, and the data are plotted as in Figure 14-1. The diamonds on the positive side of the

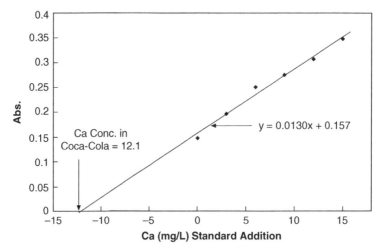

Figure 14-1. Results from a standard addition experiment in the determination of Ca in Coca-Cola.

x axis are standard concentrations that have been added to the sample. These should result in a line well above the origin (0,0) of the plot. The line is extrapolated back to a y value of zero to determine the concentration of calcium in your diluted sample. Finally, adjusting for the dilution factors that you used to make up your sample allows you to calculate the concentration of calcium in your original sample. Note that the distance from the origin to your highest standard should be of similar or less distance from the origin to your sample concentration. Thus, by using this approach we have overcome the viscosity effects and most other interferences.

The *releasing agent* is easier to understand and addresses the fact that calcium will bind to phosphate as it dries in the air–acetylene flame and therefore will not be present in its requisite form, as a ground-state gaseous atom. We use the periodicity of the elements to overcome this problem. Strontium, an element in the same group as calcium which therefore behaves much like calcium, preferentially binds to phosphate in the flame and releases calcium to form free gaseous atoms. This preferential bonding is confirmed by the much higher formation constant for strontium phosphate than for calcium phosphate. Thus, by adding another metal to each solution (standards and samples) we can overcome the dramatic effect of having phosphate in the samples. Similar approaches are available for other elements both toxic and nontoxic. Remember that if you are involved with the disposal or treatment of hazardous waste, you will want to have an accurate measurement of how much toxin is in your waste sample.

REFERENCES

Harris, D. C., *Quantitative Chemical Analysis*. 5th ed., W. H. Freeman, New York, 1999.

LeGrega, M. D., P. L. Buckingham, and J. C. Evans, *Hazardous Waste Management*, McGraw-Hill, New York, 1994.

Skoog, D. A., F. J. Holler, and T. A. Nieman, *Principles of Instrumental Analysis*, 5th ed., Harcourt Brace College Publishing, Philadelphia, 1998.

IN THE LABORATORY

The goals of this series of procedures are (1) to show that many elements and compounds can be analyzed by more than one technique and (2) to illustrate the nature of one complex sample matrix. Although the samples are not actually toxic, you should treat the sample as though it is toxic; your instructor will observe your laboratory technique, and if you handle the sample improperly, a skull and crossbones will be placed at your laboratory station.

Some of the techniques you will use are more direct and simple; others are more involved. It should be no surprise that when a single sample is analyzed by all of these techniques that the resulting concentrations do not always agree. In these labs you will measure the concentration of calcium in beverages by flame atomic absorption spectrophotometer (using external standard and standard addition techniques as well as matrix modifiers), by EDTA titration (an applied review of quantitative analysis), and using a calcium ion–specific electrode (another review of quantitative analysis), and it will be your task to decide which technique is best for your sample (by your definition or by your instructor's).

The order in which you do these procedures is not important, and they will be assigned randomly so that no two groups are doing the same lab at the same time. The most difficult step will be your first calcium determination, since you will not yet have an idea what dilution to make to analyze your first sample. However, after your first determination you will have an estimate of the concentration that can be used for the remainder of your samples dilutions and techniques. The laboratory techniques are:

- *Procedure I:* Determination of Ca Using Atomic Absorption Spectroscopy and External Standards
- *Procedure II:* Determination of Ca Using Atomic Absorption Spectroscopy, External Standards, and a Releasing Agent
- *Procedure III:* Determination of Ca Using Atomic Absorption Spectroscopy and the Standard Addition Technique with and a Releasing Agent
- *Procedure IV:* Determination of Ca Using Atomic Absorption Spectroscopy and the Standard Addition Technique without a Releasing Agent
- *Procedure V:* Determination of Ca Using the EDTA Titration
- *Procedure VI:* Determination of Ca Using Atomic Absorption Spectroscopy and Ion-Specific Electrodes
- *Procedures VII:* Additional Procedure

At the beginning of the first lab you will have to remove the carbonation by setting up a vacuum system. Also remember to sign in each logbook when you use an instrument.

Note: For each technique (except the standard addition), analyze the unknown sample five times in order to perform your statistical analysis.

USEFUL REFERENCES

Articles in the *Journal of Chemical Education*. Index search (an online searchable database) at `http://www.jchemed.chem.wisc.edu/Journal/Search/index.html`.

Ebdon, L., E. H. Evans, A. S. Fisher, and S. J. Hill, *An Introduction to Analytical Spectrometry*, John Wiley, New York, 1998.

Harris, D. C., *Quantitative Chemical Analysis*. 5th ed., W. H. Freeman, New York, 1999.

Skoog, D. A., F. J. Holler, and T. A. Nieman, *Principles of Instrumental Analysis*, 5th ed., Harcourt Brace College Publishing, Philadelphia, 1998.

Van Loon, J. C., *Analytical Atomic Absorption Spectroscopy: Selected Methods*, Academic Press, New York, 1980.

Before beginning any procedures using the AAS unit, you are expected to learn about the instrument and analytical methods by reading the relevant sections in your textbook. Some further reading will be made available from the instrument manuals. There are also guidelines for each instrument for startup and shutdown. Follow these closely!

PROCEDURE I

Determination of Ca Using Atomic Absorption Spectroscopy and External Standards

The goals of this experiment are (1) to refine your ability to make reference standards (Ca), (2) to learn to use the atomic absorption spectroscopy system using external standard calibration, (3) to determine the linear range for a set of Ca standards, and (4) to determine the concentration of Ca in an unknown sample (analyze the unknown at least five times).

Plan ahead and understand this procedure completely before you come to lab. For a review of FAAS, refer to Chapters 8 and 9 of Skoog et al. (1998). Prepare all solutions before using the FAAS unit.

External Standard Calibration Method. This is the normal way of using a calibration curve; you make a set of standards, measure the instrument's response to the standards and unknowns, make a calibration plot using linear least squares analysis (LLS; Chapter 2) (or use the automated calibration software with your instrument) and use the instrument response to estimate the concentration in your unknown samples. I suggest analyzing your standards from low to high concentration, making a blank measurement before (between) *each* standard. Repeat the standard and blank measurements twice. This will give you 15 to 20 blank measurements that you will need to determine the noise level and your minimum detection limit (equations for these are contained in the spreadsheet from Chapter 2).

1. Make a set of Ca standards (each standard should contain 1% concentrated ultrapure HNO_3 in the final solution). Calcium concentrations in the final solutions should be 0, 0.5, 1, 5, 10, 15, 25, and 50 mg of Ca per liter. (Your particular instrument may have different detection limits and linearity ranges from those used to develop this experiment. Consult your instructor for details on your instrument.) Note that some of these standards will be below the detection limit while others may be above the limit of linearity.
2. Make five dilutions of the unknown sample.
3. Set up the FAAS unit as instructed.
4. Analyze the standards and unknown samples on the AAS unit.
5. Plot the data using your LLS spreadsheet, determine the linear portion of the data, and if the unknown sample signal is in the linear range, determine the concentration of Ca. If the signal of the sample is too high, make the appropriate dilution of the sample in 1% HNO_3, and reanalyze the sample.

PROCEDURE II

Determination of Ca Using Atomic Absorption Spectroscopy, External Standards, and a Releasing Agent

The goals of this experiment are (1) to refine your ability to make reference standards (Ca), (2) to learn the use of releasing agent in FAAS, (3) to learn to use the atomic absorption spectroscopy system, (4) to determine the linear range for a set of Ca standards, and (5) to determine the concentration of Ca in an unknown sample (analyze the unknown at least five times).

This procedure is identical to Procedure I, except that you will have to add a releasing agent (Sr) to every solution. The final concentration of Sr in all of your standards and samples should be 1000 mg/L. To achieve this, you will have to make a more concentrated Sr solution and add a small but consistent volume of this concentrated solution to your standards and samples.

Note: Do all calculations for dilutions and preparing solutions before you come to lab or you will be very late leaving on lab day.

Again, analyze your standards from low to high concentration and make a blank measurement before (between) *each* standard. Repeat the measurement of standards and blanks twice. This will give you 15 to 20 blank measurements that you will need to determine the noise level and your minimum detection limit (equations for these are contained in your spreadsheet from Chapter 2).

1. Make a stock solution of $Sr(NO_3)$ at a concentration that will serve to meet the requirements below. Check with your instructor before you make the solutions to ensure that you have the calculations correct.
2. Make a set of Ca standards (each standard should contain 1% concentrated ultrapure HNO_3 in the final solution) and 1000 mg of Sr per liter. Calcium concentrations in the final solutions should be 0, 0.5, 1, 5, 10, 15, 25, and 50 mg of Ca per liter. (Your particular instrument may have different detection limits and linearity ranges from the one used to develop this experiment. Consult the instructor for details on your instrument.) Note that some of these standards will be below the detection limit, whereas others may be above the limit of linearity.
3. Make five dilutions of the unknown sample and add Sr to a level of 1000 mg/L.
4. Set up the AAS unit as instructed.
5. Analyze the standards and unknown samples on the AAS unit.
6. Plot the data using your LLS spreadsheet, determine the linear portion of the data, and if the unknown sample signal is in the linear range, determine the concentration of Ca. If the signal of the sample is too high, make the appropriate dilution of the sample in 1% HNO_3, add Sr to 1000 mg/L, and reanalyze the sample.

PROCEDURE III

Determination of Ca Using Atomic Absorption Spectroscopy and the Standard Addition Technique with a Releasing Agent

The goals of this experiment are (1) to refine your ability to make reference standards (Ca), (2) to learn to use the atomic absorption spectroscopy system, (3) to learn the standard addition technique, (4) to learn one technique for overcoming interferences (releasing agents), and (5) to determine the concentration of Ca in an unknown sample.

Plan ahead and understand this procedure completely before you come to lab. Prepare all solutions before using the FAAS unit.

Note: Do all calculations for dilutions and preparing solutions before you come to lab or you will be very late leaving on lab day.

Standard Addition Calibration Method. Here we are concerned with viscosity effects from the corn syrup in your hazardous waste sample. We also evaluate the affect of adding a releasing agent (Sr). You should understand completely why you are adding this before you come to lab.

1. Make a stock solution of $Sr(NO_3)$ at a concentration that will serve to meet the requirements below. Check with your instructor before you make the solutions to ensure that you have the calculations correct.
2. Make a set of standards and samples containing known amounts of Ca (standard) and Sr (at 1000 mg/L in the final solution). Calcium concentrations in the final solutions should be 0, 0.5, 1, 5, 10, 15, 25, and 50 mg of Ca per liter. (Your particular instrument may have different detection limits and linearity ranges than the one used to develop this experiment. Consult your instructor for details on your instrument.) When you make these solutions, I suggest making the samples in 25-, 50-, or 100-mL volumetric flasks [i.e., to each volumetric flask (a) add an exact and equal volume of sample, based on one of your other experimental results; (b) add concentrated HNO_3 to yield 1%; (c) add a volume of $SrNO_3$ solution that will give you 1000 mg of Sr per liter; and (d) fill the flask with distilled water to the mark.] Note that you need your sample concentration (on the $-x$ axis) to be within the range of your sample plus standard concentrations (on the $+x$ axis).
3. Analyze the standards and samples on the FAAS unit.
4. Make sure that the data set is linear. If it is not, consult your laboratory instructor before you throw away your solutions.
5. Plot the data using your LLS spreadsheet, determine the linear portion of the data, and if the unknown sample signal is in the linear range, determine the concentration of Ca. If the signal of the sample is too high, make the appropriate dilution of the sample in 1% HNO_3, add 1000 mg Sr per liter, and reanalyze the sample.

PROCEDURE IV

Determination of Ca Using Atomic Absorption Spectroscopy and the Standard Addition Technique without a Releasing Agent

The goals of this experiment are (1) to refine your ability to make reference standards (Ca), (2) to learn to use the atomic absorption spectroscopy system, (3) to learn the standard addition technique, (4) to learn one technique for overcoming interferences, (5) to determine the concentration of Ca in an unknown sample.

This procedure is identical to Procedure III, but you will not be using Sr as a releasing agent. Delete all reference to it and complete Procedure III.

PROCEDURE V

Determination of Ca Using the EDTA Titration

The goals of this experiment are (1) to refine your ability to make reference standards (Ca) and dilutions, (2) to review and refine your titration skills, (3) to review or learn the details of a complicated EDTA titration, and (4) to determine the concentration of Ca in an unknown sample.

Plan ahead and outline a procedure completely before you come to lab. In this procedure you may use Eriochrome Black T, but a better indicator is solid hydroxynaphthol blue.

Use your knowledge from quantitative analysis to conduct this experiment. Note that you may have to dilute your sample (and possibly the EDTA) to dilute the food coloring, which may interfere with the endpoint to obtain an acceptable detection limit. It will also be important for you to review exactly what the EDTA titration is measuring as compared to the other procedures in this set of laboratory exercises.

1. Pipet a sample of your unknown into a 250-mL flask. You will have to determine the initial dilution of the sample and EDTA titrant. The beginning of the procedure will be highly dependent on a trial-and-error approach, and there is more than one correct way of completing this procedure. To each sample aliquot that you titrate (below), add 3 mL of the pH 10 buffer solution and 30 drops of 50% by weight NaOH, swirl for 2 minutes, and add a small scoop (about 0.1 g) of hydroxynaphthol blue (or 6 drops of a Eriochrome Black T indicator solution). (*Note*: Your sample is naturally acidic, so you may need to add more than 30 drops of NaOH. Check the pH to ensure that it is at or above 10.)

2. After you have determined the best dilutions of the sample and EDTA titrant to use, complete at least three sample titrations to find the amount of Ca^{2+} in your unknown sample. Note that you may need to add deionized water to your flask to give a sufficient volume for your titration.

3. Titrate the Ca determinations carefully. After reaching the blue endpoint, allow each sample to sit for 5 minutes, with occasional swirling, so that any $Ca(OH)_2$ precipitate can redissolve (if this occurs, the solution will be red or pink). Then titrate back to the blue endpoint. It is always best to perform a blank titration on deionized water to serve as an endpoint check, but note that your sample has a background color.

4. Calculate the total Ca concentration in your original sample (1 mole of Ca^{2+} binds with 1 mole of EDTA).

PROCEDURE VI

Determination of Ca Using Atomic Absorption Spectroscopy and Ion-Specific Electrodes

The goals of this experiment are (1) to refine your ability to make reference standards (Ca) and dilutions, (2) to review/learn the details of ion-specific electrodes, and (3) to determine the concentration of Ca in an unknown sample.

Plan ahead and outline a procedure completely before you come to lab. This will involve reading the manual for your Ca electrode. You should also review solid-state electrodes in a quantitative analysis textbook.

Follow the instructions in the electrode manual, and make an external calibration curve to check the slope of the line to ensure that the electrode is functioning properly and for your LLS analysis. You may also choose to analyze a set of samples using the standard addition technique.

PROCEDURE VII

Additional Procedure

If you have an inductively coupled plasma (ICP) instrument and a voltametry setup, you can also measure the Ca concentration using these techniques.

ASSIGNMENT

What do you turn in? One of the goals of this lab manual/course is not only to teach you proper methods for analyzing samples, but also to teach you to communicate your results effectively. Apart from lab notebooks and lab reports that you will complete for this and other labs, in this lab exercise you will do something a little more involved. After completion of all procedures, you are to compile the methods and results and write a journal article suitable for publication in the *Journal of Analytical Chemistry*. The theme of your article will be comparing analytical techniques for calcium analysis of complex aqueous samples. You must obtain the "Instructions to Authors" for the journal from the library or Internet and follow proper scientific writing guidelines (refer to the *ACS Style Manual* on reserve in the library). Remember that in your lab reports you write down meticulous lab methods, but you will not be able to do this in your journal article (if you did this, the article would be 50 pages long!). You must decide the fine line between too little and too much information. The best and perhaps the only way to do this is to review several articles in the journal (perhaps two or three on AAS, two or three on titration techniques, and two or three on ion-specific electrodes). Note that you *must* also do a literature search on your topic and include the results in the introduction. For the introduction you can begin the article from a hazardous waste or analytical standpoint. Your article should be no longer than 25 typed double-spaced pages, including text, figures, tables, and references. In your discussion and conclusions section, *defend* which method(s) is(are) most accurate for determining Ca in your sample.

DATA COLLECTION SHEET

DATA COLLECTION SHEET

DATA COLLECTION SHEET

DATA COLLECTION SHEET

DATA COLLECTION SHEET

DATA COLLECTION SHEET

DATA COLLECTION SHEET

DATA COLLECTION SHEET

DATA COLLECTION SHEET

DATA COLLECTION SHEET

DATA COLLECTION SHEET

DATA COLLECTION SHEET

DATA COLLECTION SHEET

15

REDUCTION OF SUBSTITUTED NITROBENZENES BY ANAEROBIC HUMIC ACID SOLUTIONS

Purpose: This laboratory experiment serves as a capstone exercise for an environmental chemistry course and includes concepts of solution preparation, pH buffers, E_H buffers and solutions, organic reaction mechanisms, reaction kinetics, and instrumental analysis (HPLC or GC). In this exercise students use a simulated hazardous waste sample from a landfill and study the first-order degradation of substituted nitrobenzenes to anilines.

BACKGROUND

Biotic (microbially mediated) and abiotic (chemical mediated with no microbial involvement) pollutant transformation reactions have long been recognized as important in determining the life-cycle toxicity of a compound. Both anaerobic and aerobic transformations can occur. Aerobic transformations include the partial degradation of an organic pollutant to by-products as well as complete mineralization to carbon dioxide. Anaerobic transformations include dehalogenations, nitro reductions, dealkylations, azo-linkage reductions, and sulfoxide and sulfone reductions. Two excellent reviews of these abiotic, anaerobic reactions can be found in Macalady et al. (1986) and Schwarzenbach and Gschwend (1990).

Environmental Laboratory Exercises for Instrumental Analysis and Environmental Chemistry
By Frank M. Dunnivant
ISBN 0-471-48856-9   Copyright © 2004 John Wiley & Sons, Inc.

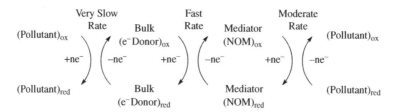

Figure 15-1. Conceptual representation of the electron shuttle system. (Modified from Glass, 1972.)

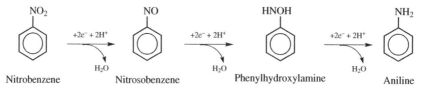

Figure 15-2. Reaction mechanism for the reduction of nitrobenzene to aniline. (From March, 1985.)

The basic abiotic reaction used in this exercise to illustrate anaerobic degradation (Figure 15-1) is similar to the transformation of pollutants by microorganisms in the environment. However, in the abiotic reactions, natural organic matter (NOM) takes the place of microbes to shuttle the electrons from the bulk electron donor (in our case, hydrogen sulfide) to the pollutant (in our case, a nitro group on substituted benzenes). In this reaction, the reaction rate is relatively slow between the pollutant and the bulk electron donor. However, the bulk electron donor reduces the natural organic matter rapidly, which in turn reduces the organic pollutant rapidly. The sequence of reductions for nitrobenzene and substituted nitrobenzenes is shown in Figure 15-2.

Before the laboratory exercise is attempted, students should read and discuss (in lecture class) the papers of Macalady et al. (1986), Schwarzenbach et al. (1990), and Dunnivant et al. (1992). In the laboratory experiment following the lecture, students study one or all of the chemical reduction experiments presented here.

ACKNOWLEDGMENT

Mark-Cody Reynolds (Whitman College, Class of 2004) collected all of the experimental data for this exercise.

REFERENCES

Dunnivant, F. M., R. P. Schwarzenbach, and D. L. Macalady, *Environ. Sci. Technol.*, **26**, 2133–2141 (1992).

Glass, B. L., *Agric. Food Chem.,* **20**, 324 (1972).

Macalady, D. L., P. G. Tratnyek, and T. J. Grundl, *J. Contamin. Hydrol.,* **1**, 1–28 (1986).

March, J., *Advanced Organic Chemistry: Reductions, Mechanisms, and Structure,* 3rd ed., Wiley, New York, 1985.

Schwarzenbach, R. P., R. Stierli, K. Lanz, and J. Zeyer, *Environ. Sci. Technol.,* **24**, 1566 (1990).

Schwarzenbach, R. P., and P. M. Gschwend, Chemical transformations of organic pollutants in the aquatic environment, in *Aquatic Chemical Kinetics,* W. Stumm (ed.), Wiley, New York, 1990.

IN THE LABORATORY

All solutions for this laboratory will be prepared by the laboratory assistant. This is a complicated experiment, and you should proceed carefully through the procedures. The laboratory exercise will take at least two weeks to complete. During the first laboratory period, you will make your experimental solution for the degradation. In the time between the first laboratory period and the second, you will be assigned times to come into the laboratory, take samples, and extract them into organic solvent to stop the reduction reaction. During the second laboratory period, you will analyze the samples on either the GC or HPLC.

Safety Precautions

- Although this experiment presents no unusual hazards, standard precautions should be used in handling organic solvents and in disposing of organic wastes. Dispose of organic wastes in an organic waste container.
- Avoid inhaling the H_2S vapors by always using the fume hood for all manipulations involving the stock solution and samples.
- Use gloves as needed when handling organic compounds.

Chemicals and Solutions

- 0.100 M solutions of all substituted nitrobenzenes.
- ACS-grade Na_2S used as the bulk electron donor for the reduction reaction. A stock solution of about 0.50 M S^{-2} is prepared and standardized with thiosulfate solution. The volume recommended is 100 mL stored in a 125-to 150-mL septum-capped serum vial.
- The iodometric titration method is used to standardize the thiosulfate titrant for measuring the concentration of sulfide in the stock solution. To accomplish this, the following chemicals are required: (1) About a 0.025 M solution of sodium thiosulfate; (2) a standard potassium biiodate solution (0.00210 M); and (3) KI crystals, concentrated H_2SO_4, and starch indicator.
- The pH of the system is controlled with a HEPES [N-(2-hydroxyethyl)piperazine-N'-2 ethane-sulfonic acid] buffer solution. A 1.00 L stock solution (0.200 M) is required.
- The natural organic matter used in this experiment is a specially prepared Fluka solution. The standardization procedure is described in the notes to the instructor.
- *For GC analysis*: Ethyl acetate is used for extraction of samples. Ethyl acetate is spiked with 100 μM unsubstituted nitrobenzene, an internal standard (for GC analysis only). A sample volume (from your experimental solution) of 0.500 mL is added to a 2.5-mL autosample vial. Ethyl acetate (0.500 mL) is added to the vial and shaken for 1 minute to extract the

nitrobenzene compounds into solution. The resulting ethyl acetate layer is withdrawn with a Pasteur pipet and placed in another autosampler vial for GC or HPLC analysis.

- *GC standards*: A range of standards in ethyl acetate are required, ranging from 5 to 100 μM, depending on the detection limit of your instrument. All standards and sample extracts should contain 100 μM unsubstituted nitrobenzene as an internal standard.

- *For HPLC analysis*: Ethyl acetate is used for extraction of samples. Ethyl acetate can be spiked with 100 μM unsubstituted nitrobenzene to act as an internal standard, but in HPLC analysis there is no real advantage in this (in fact, it may interfere with the resolution of your analyte nitrobenzene). A sample volume of 0.500 mL is added to a 2.5-mL autosample vial. Ethyl acetate (0.500 mL) is added to the vial and shaken for 1 minute to extract the nitrobenzene compounds into solution. The resulting ethyl acetate layer is injected into the HPLC system using a six-port valve.

- *HPLC standards*: A range of standards in ethyl acetate is required, from 5 to 100 μM, depending on the detection limit of your instrument. Injection volumes can range from 10 to 25 mL.

Equipment

- *GC analysis*: a capillary column GC with a flame ionization detector. The capillary column should be an HP-1, 30 m by 0.320 mm outside diameter, with a 0.25-μm film thickness.

- *HPLC analysis*: an HPLC equipped with a high-pressure pump, UV–Visible detector, six-port sampling valve, and C-18 HPLC column (10 cm by 4.6 mm). The mobile phase used was 0.01 M hydroxylamine hydrochloride buffer (pH 6.0) in methanol/water (typically 3:2 v/v). The flow rate was 1.0 mL/min and the injection volume was 6 to 10 μL.

PROCEDURE

A basic procedure for study of the first-order degradation process is given below. For the other experiments (pH dependence, dependence on substitute pattern, dependence on humic acid concentration) you will have to modify the procedures slightly.

Preparation of Experimental Solutions in Serum Vials

In this procedure you will make a solution containing H_2S, HEPES pH 7.2 buffer, 3-chloronitrobenzene, HCl to neutralize the basic nature of the S^{2-}, deionized water, and Fluka humic acid. Blank solutions should also be taken through the procedure and are identical to the experimental solution but without humic acid. You should keep a record of the volumes of each solution added since you will add deionized water in step 3 to bring the total volume to 40 mL.

1. To a 50- to 75-mL serum vial, add 20.0 mL of the 0.10 M pH 7.2 HEPES buffer solution.

2. Add an appropriate volume of the filtered Fluka humic acid solution to obtain the desired NOM concentration (and the desired rate). To obtain a concentration of 26 mg/L, you will need to add approximately 1.0 mL of stock NOM solution. Use $M_1V_1 = M_2V_2$ to determine the needed volume of your stock NOM solution. Your stock Fluka humic acid solution should be at a concentration of 1000 mg/L.

3. Add 0.300 mL of 1 M HCl for the pH 7.2 solutions (this will neutralize the basic nature of the S^{2-}).

4. Add sufficient deionized water to bring the solution to a standard volume of 40 mL after addition of the following solutions.

5. Purge the solution of atmospheric oxygen by attaching a low-pressure N_2 source to Tygon tubing and then to a syringe needle. Place an additional needle in the septum (but not connected to the N_2 source) to allow the atmospheric oxygen and added N_2 to exit the system. Purge the serum vial for at least 5 minutes.

6. With a syringe and needle, add a volume of the calibrated stock Na_2S solution to obtain 5 mM S^{2-} and let the solution sit *overnight* to equilibrate with the natural organic matter (NOM). Depending on the concentration of your stock solution, the volume of 0.5 M S^{2-} should be approximately 0.40 mL.

7. With a syringe and needle, add sufficient 0.100 M 3-chloronitrobenzene (or other substituted nitrobenzene) to obtain a final concentration of 100 μM nitrobenzene in the serum vial. Depending on your stock solution concentration, the volume will be approximately 40 μL.

8. Shake the solution and sample immediately for an initial concentration measurement of your analyte. Sample by filling a 0.50-mL glass syringe

with nitrogen gas (to avoid the introduction of atmospheric oxygen), inserting the needle through the septum, adding the nitrogen, and filling the syringe with solution. Remove the syringe from the serum vial, remove all gas bubbles, and adjust the volume to 0.50 mL. Add the 0.50 mL to a 2-mL vial containing 0.500 mL of ethyl acetate solution. (If you are analyzing your samples by GC–flame ionization detector, you will need to have unsubstituted nitrobenzene in the ethyl acetate at a concentration of 100 μM).

9. After adding your 0.50-mL sample to each vial for analysis, seal it, shake it rigorously, open the vial to add more oxygen, and repeat once more to ensure that the reduction reaction is stopped (the oxygen will oxidize the sulfide and stop the reduction process).

10. After the initial sample, take samples of your solutions at timed intervals based on the humic acid concentration and the expected rate. Follow steps 8 and 9 for these as well. You should collect approximately 10 data points, ranging from time zero through three half-lifes of your reaction.

Analyze the ethyl acetate layer for substituted nitrobenzene using a gas chromatograph or high-performance liquid chromatograph.

GC Conditions

- Temperatures:
    Front inlet = 250 °C
    Detector = 250 °C
- Inject 1 μL of sample
- Flame ionization detector (hydrogen–air flame)
- He carrier gas
- Column: Agilent Technologies HP-5, 30.0 m by 320 mm by 0.25 μm inside diameter
- Temperature program:
    Initial temperature: 75°C for 10.00 minutes
    Ramp 1:   10.00°C/min to 135°C, hold for 17.0 minutes
    Ramp 2:   20.0°C/min to 230°C, hold for 5 minutes for cleaning

HPLC Conditions. C-18 HPLC column (10 cm by 4.6 mm). The mobile phase used was 0.01 M hydroxylamine hydrochloride buffer (pH 6.0) in methanol/water (typically 3:2 v/v). The flow rate was 1.0 mL/min and the injection volume was 6 to 10 μL.

Waste Disposal

All solutions should be disposed of in an organic waste container.

ASSIGNMENT

To be determined by your laboratory instructor, depending on whether you conduct the basic nitro-reduction experiment or another experiment related to this reduction.

ADVANCED STUDY ASSIGNMENT

1. List and give anaerobic reduction reactions for three important pollutants.
2. How is the bulk electron donor involved in these reactions?
3. Explain how you will determine the first-order degradation rate from your data set.

DATA COLLECTION SHEET

DATA COLLECTION SHEET

DATA COLLECTION SHEET

DATA COLLECTION SHEET

DATA COLLECTION SHEET

PART 5

EXPERIMENTS FOR SEDIMENT AND SOIL SAMPLES

16

SOXHLET EXTRACTION AND ANALYSIS OF A SOIL OR SEDIMENT SAMPLE CONTAMINATED WITH *n*-PENTADECANE

Purpose: To use the Soxhlet extraction apparatus to extract a hydrocarbon pollutant from a soil or sediment sample

To learn the finer points of analyte recovery in trace organic analysis

To learn to use internal standards to quantify analyte recovery

BACKGROUND

One of the most challenging aspects of environmental chemistry is the incorporation of analytical chemistry into environmental monitoring. In this lab we illustrate some of the finer points of environmental monitoring. Soils and sediments around the world are contaminated with a variety of inorganic, organic, and radioactive pollutants. This laboratory exercise concentrates on organic contamination that can occur from industrial spills and leaks from storage tanks. Even changing the oil in your car or spilling fuel at a gasoline station can result in soil contamination that is difficult to clean up (*remediate*). Soil contamination can be mild to severe, ranging from part-per-million levels to percentage levels. On the other hand, lake and river sediment contamination is usually at low concentrations (parts per billion or parts per million). This contamination results from smaller gasoline or industrial spills that enter a water body and adsorb to the surface or interior of the sediment particles. Sediment contamination is slightly more difficult to document since the

Environmental Laboratory Exercises for Instrumental Analysis and Environmental Chemistry
By Frank M. Dunnivant
ISBN 0-471-48856-9   Copyright © 2004 John Wiley & Sons, Inc.

contaminant concentration can vary greatly within a water body, and more care must be taken to collect and analyze low pollutant concentrations accurately.

The analysis of petroleum hydrocarbons is a recurring theme in this laboratory manual. This is because they are practically ubiquitous in the environment. The EPA estimates that in the United States there are approximately 705,000 underground storage tanks (USTs) that store petroleum or hazardous substances that can harm the environment if released (U.S. EPA, 2003). As of September 2001, over 418,000 UST releases have been documented. During this time, 268,000 contaminated sites have been cleaned up, but there are about 150,000 sites remaining to be remediated (U.S. EPA, 2003). The scale of the UST problem has led the EPA to create a major special program to address this problem.

METHODS OF EXTRACTION

A number of extraction methods have been developed for recovering organic pollutants from soil and sediment samples. These include shake extraction methods (Cotterill, 1980), ultrasonication (Johnson and Starr, 1972; Dunnivant and Elzerman, 1988), heated solvent extraction (Dionex Corp, Inc.), steam distillation (Swackhamer, 1981), and Soxhlet extraction (the subject of this experiment; Poinke et al., 1968; Fifield and Haines, 2000; Perez-Bendito and Rubio, 2001). *Shake extraction methods* involve placing the soil or sediment sample in a sealed flask containing an organic solvent (which is usually miscible with water, since most samples are extracted field-wet, with no drying). The flask is placed on a shaker table and mixed overnight to extract the contaminants into the organic solvent. Some procedures call for replacing the solvent and repeating the shaking for another 24-hour period. Shake methods have been found to be the least effective at extracting contaminants from soil and sediment samples.

Ultrasonication methods involve placing the soil or sediment sample in a small beaker or vial containing organic solvent and disrupting the sample with sonic energy delivered through a probe. Again, a water-miscible solvent is used because the sample is usually extracted field-wet. Sonication methods are highly effective at breaking up the sample aggregates and extracting the contaminants, but are slightly less effective than the Soxhlet extraction method. Dionex has developed a heated solvent extraction system in which the soil or sediment sample is placed in a tube and heated organic solvent is passed through the sample. This method has two advantages: (1) the heated solvent increases diffusion of contaminants out of the sample, and (2) the system is automated, so that several samples can be extracted at one time. This procedure is highly effective at extracting contaminants, but the apparatus is expensive.

Steam distillation is a technique in which an aqueous suspension of the sample is placed in a flask and steam is used to remove the semivolatile contaminants. Volatilized contaminants are recovered in a thimble containing organic solvent. This procedure appears to be slightly less effective than sonication and Soxhlet extraction. By far the most rigorous, time-consuming, and effective method of extraction is by use of the *Soxhlet apparatus*. This technique has been tested for decades and almost always yields the highest recovery of contaminants. We illustrate and use this technique in this laboratory exercise.

THEORY

A typical Soxhlet apparatus is shown in Figure 16-1. The three main components are the condenser, which cools the solvent vapor into a liquid that contacts the sample, the contact/extraction chamber, which holds the fiber thimble containing the sample; and the boiling flask, which holds the solvent and extracted analytes. The boiling flask is heated with a heating mantle. As the solvent is heated, it refluxes and vapor rises through the transfer tube on the far right side of the contact/extraction chamber. The vapor continues up into the condenser, where it is cooled and drips onto the top of the thimble. As the solvent contacts the soil or sediment sample, it extracts the pollutants into the solvent phase, which collects in the extraction chamber. As the solvent level in the extraction chamber increases, it eventually reaches the top of the recycle tube (the curved tube between the vapor transfer tube and the extraction chamber). The recycle tube transfers the pollutant-laden solvent back into the boiling flask, where the analyte remaining in the hot solvent (if less volatile than the solvent) is refluxed back into the extraction chamber. Usually, the heating level is adjusted so that it takes from 10 to 20 minutes for the extraction chamber to fill and empty. Soxhlet extractions can be conducted for 8 to 24 hours, depending on the difficulty of extraction. All Soxhlet extractions using organic solvent should be conducted in a fume hood.

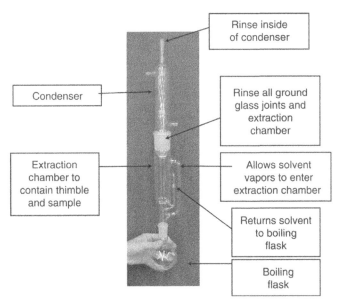

Figure 16-1. Soxhlet extraction apparatus.

If we consider only one solvent–soil contact, the Soxhlet technique uses basically a shaker method extraction. Using this approach, we can define a distribution ratio D, which describes the equilibrium analyte concentration C_a between the soil and the solvent phases:

$$D = \frac{[C_a]_{\text{soil}}}{[C_a]_{\text{solvent}}}$$

The extraction efficiency is given by

$$E = \frac{100D}{D + V_{\text{solvent}}/V_{\text{soil}}}$$

When D is greater than 100, a single equilibrium extraction will quantitatively extract virtually all of the analyte into the solvent phase. However, since we allow only 10 to 20 minutes per cycle, we rarely have equilibrium conditions after one extraction. So we reflux the Soxhlet over a longer period and maximize the concentration gradient into the solvent phase by replacing pure solvent in the extraction chamber. Also, since we rarely know the D value for the soil we are extracting, the exhaustive extraction (24 hours) hopefully ensures that we quantitatively extract all or most of the analyte from the soil.

REFERENCES

Cotterill, E. G., *Pestic. Sci.*, **11**, 23–28 (1980).

Dunnivant, F. M. and A. W. Elzerman, *J. Assoc. Offic. Anal. Chem.*, **71**, 551–556 (1988).

Fifield, F. W. and P. J. Haines, *Environmental Analytical Chemistry*, 2nd ed., Blackwell Science, London, 2000.

Johnson, R. E. and R. I. Starr, *J. Agric. Food Chem.*, **20**, 48–51 (1972).

Perez-Bendito, D. and S. Rubio, *Environmental Analytical Chemistry*, Elsevier, New York, 2001.

Poinke, H. B., G. Chesters, and D. E. Armstrong. Extraction of chlorinated hydrocarbon insecticides from soils. *Agronomy J.* **60**, 289–292 (1968).

Swackhamer, D. L., Master's thesis, Water Chemistry Program, University of Wisconsin, Madison, WI, 1981.

U.S. EPA, `http://www.epa.gov`, 2003.

IN THE LABORATORY

We will be extracting a sand sample that has been contaminated with a known mass of *n*-pentadecane (C-15). This compound does not usually occur as an isolated contaminant but is present in mixtures of hydrocarbons. We will be using *n*-pentadecane as a surrogate contaminant for any petroleum hydrocarbon, since all petroleum hydrocarbons can be extracted and analyzed as shown in this experiment. The extraction method that we use is the Soxhlet technique, coupled with internal standard additions and gas chromatography. As noted earlier, Soxhlet extraction is the most common and accepted form of extraction of organic contaminants from soil and sediment samples. It has long been recognized by EPA as the best method of extraction for these types of samples. But it is labor intensive and requires more time than some of the other techniques discussed.

Source of Error and Internal Standards

There are several steps (addition of contaminated sand to vessels, addition of internal standards, transfer of extraction fluids, etc.) in this lab where good lab technique is of extreme importance. Poor quantitative transfer in these steps will result in low recovery of your analyte (C-15) and/or internal standards. We will evaluate your lab technique by the use of internal standards, although these are normally used in the real world to correct for poor extraction techniques and unavoidable analysis errors. Two types of internal standards are used. First, an internal standard is used to check the recovery of C-15 in the Soxhlet extraction technique, since some of it may volatilize during extraction. Note that this recovery standard evaluates primarily your laboratory technique, not the ability of the Soxhlet to extract pollutants from soil particles. These are two very different concepts. To test the extraction efficiency of Soxhlet extraction we would need to know the exact level of contamination or spike the soil or sediment and let it mix for months to years for complete sorption equilibrium to be achieved. For our purposes (to check your laboratory technique), you will add a known mass of a similar compound (C-17) to the Soxhlet apparatus and extract it in the process of extracting the C-15 from your sample. Any losses from volatilization during extraction or from transferring the solvent containing the analytes (C-15 and C-17) should be accounted for with the C-17 recovery standard.

There are several common losses of analyte during the Soxhlet extraction procedure. A few simple ones include improper weighing of soil or sediment, the use of plastic materials that will sorb organic analytes from a sample and contaminate a sample with phthalates, and inaccurate addition of the C-17 internal standard. Less obvious sources of error are also common. After the extraction process but before the Soxhlet apparatus is disassembled, the condenser should be rinsed with approximately 5 to 10 mL of pure extraction solvent (in our case, methylene chloride) (refer to Figure 16-1). During the isolation and removal of the extraction solvent, all ground-glass joints should be rinsed with a small volume of pure extraction solvent. Finally, the boiling flask and boiling beads should be rinsed with pure extraction solvent. All of these rinses should be added to the volumetric flask containing the extraction solvent. Note that every drop you lose or spill will contain C-15, and your recovery standards and can result in low analyte recoveries.

The second internal standard (C-13) will account for analyte losses during injection into the gas chromatograph. As the liquid sample extract is added to the injector of the gas chromatograph, the solvent expands greatly and increases the pressure in the injection chamber. When you withdraw the needle, some of the volatilized injected solvent will escape through the septum. Unfortunately, the amount of solvent (and analyte) that escapes is not consistent between injections, and we need some way of knowing just how much C-15 and C-17 is lost from the injection system. By adding a known mass of our second internal standard (C-13) to each standard and sampling immediately before analysis on the GC, we will have a way of measuring these losses. Most modern gas chromatography data-handling stations will account for this second internal standard in any data reporting schemes.

The final task of this experiment is to convert the concentration of analyte in your extraction solvent back into the concentration in the original sample. This is completed by keeping track of the soil or sediment masses and dilution factors. An example of these calculations is contained in the Advanced Study Assignment.

PreLab Demonstrations

Soxhlet Setup. The Soxhlet extraction system will be demonstrated in class. Take careful note that all of the extractors are connected with a cooling water hose, and note that each extractor costs over $250! Note the points where analyte loss or contamination can occur:

- Insufficient precleaning of glassware (organic contaminants are always present)
- Contact with plastic and rubber materials
- Use of non-precleaned extraction thimbles
- Contamination in weighing sample and sample handling
- Loss of analyte during extraction (C-17 internal standard)
- Loss of analyte during recovery and dilution of solvent (ground-glass joints and rinsing)
- Solvent concentration step
- Solvent recovery steps

GC/FID. We will be using a capillary column gas chromatograph (GC) equipped with a flame ionization detector (FID) to analyze for *n*-pentadecane, along with the two internal standards. This instrument is designed to analyze semivolatile compounds in the part-per-million to part-per-thousand range. More specifically, the FID system only detects compounds that will burn (ionize) in a hydrogen–air flame. Your instructor will go over the setup of the instrument, how to inject samples, and how to interpret the output before you use the instrument.

Safety Precautions

- Safety glasses must be worn during this laboratory experiment.
- Most of the chemicals used in this experiment are flammable. Do not have an open flame in the laboratory.
- Soxhlet extractions should be preformed in a fume hood.
- Review material safety and data sheets (MSDSs on the hydrocarbons and for methylene chloride.
- The heating of the methylene chloride in the boiling flasks should be increased incrementally to avoid bumping of the solvent. Boiling chips *must* be used to avoid a pressure explosion of the glassware.

Chemicals and Solutions

- Neat C-13 (GMW 184.47, density $= 0.7564$ g/mL), C-15 (GMW 212.42, density $= 0.7685$ g/mL), and C-17 (GMW 240.48, density $= 0.7780$ g/mL).
- *GC calibration standards*: Add 2.00 µL (yields 15.37 mg/L), 5.00 µL (yields 38.42 mg/L), 10.00 µL (yields 76.85 mg/L), and 25.00 µL (yields 192 mg/L) of pure (neat) C-15 to separate 100-mL volumetric flasks. To each 100-mL flask add 1.00 mL of a 3026-mg/L C-13 solution (below; yields 30.26 mg/L in your standards) and 40.0 µL of a 77,800-mg/L C-17 solution (below; yields 31.12 mg/L in your standards). Use methylene chloride as the solvent to fill the flask to the 100-mL mark.
- *Spiked sand.* Each student will need approximately 35 g of sample. The sand contains minimal moisture, so we do not have to use a water-miscible solvent.
- *C-17.* Make a 1 : 10 dilution of the neat C-17 stock (yields 77.8 mg/mL or 77,800 mg/L) for addition to the sand and thimble. The final concentration of C-17 in the concentrated 1.0-mL extract will be 31.12 mg/L.
- CH_2Cl_2 for extraction: pesticide grade.
- C-13 addition to concentrated sample. Make a 3026-mg/L solution of C-13 in methylene chloride by adding 100 µL of neat C-13 to 25 mL of CH_2Cl_2. Add 10.0 µL to the 1.0-mL concentrated samples in the Kuderna–Danish thimble, which will yield a concentration of 30.26 in the 1-mL concentrated extract.

Equipment and Glassware

- Kuderna–Danish thimbles, one per Soxhlet apparatus

- Soxhlet setups (one per student pair and one blank)
- Heating mantles
- Preextracted thimbles
- 1.00-mL pipets
- 250-mL volumetric flasks
- 1.5- to 2.00-mL autosampler vials
- Three 10.0-μL syringes for adding C-13, C-15, and C-17
- Capillary column GC, DB-1 column (a variety of GC conditions will work for this separation, since the hydrocarbons separate very easily; a basic temperature program for the oven follows)

GC Conditions

- Backpressure on column: 6.30
- Carrier gas flow rate in column: 1.3 mL/min
- Linear velocity: 23 cm/s
- Initial oven temperature: 135°C
- Initial hold time: 2 minutes
- Oven program rate: 5°C/min
- Final oven temperature: 210°C, hold for 5 minutes
- Injector temperature: 230°C
- Detector temperature: 250°C
- Approximate retention times (depending on column length and coating thickness):

Elution Order	Time (min)
C-13	9.80
C-15	12.30
C-17	14.60

PROCEDURE

Week 1

Note: Always work with someone when you are handling the Soxhlet setups to avoid disaster and an expensive glassware bill!

1. Rinse your entire Soxhlet apparatus in a fume hood with clean methylene chloride.
2. Obtain a preextracted thimble (thimbles usually come precleaned from the factory).
3. Place 200 mL (no more) of pesticide-grade methylene chloride in a 250-mL boiling flask. This will leave 50 mL for rinsing. Add boiling chips to the flask. Place the flask in the heating mantle.
4. Weigh out into your thimble between 25 and 35 g of contaminated sand (note the contamination sources mentioned in prelab). Record the weight to the nearest 0.01 g.
5. Gently place the thimble in the Soxhlet cylinder so that no sand or sediment spills out into the extraction chamber.
6. Add 10.0 µL of the 1 : 10 dilution of neat C-17 to the sand/thimble (yields a C-17 concentration of 3.112 mg/L in your 250-mL volumetric flask if you assume that all of the C-17 is extracted).
7. Connect the Soxhlet cylinder and the condenser.
8. After everyone has assembled a Soxhlet setup, you will turn on the condenser water and reflux the methylene chloride for about 24 hours. The extraction chamber should fill and empty every 10 to 20 minutes.

Week 2

1. Quantitatively transfer all of the methylene chloride in your Soxhlet setup to a 250-mL volumetric flask, noting the sources of error mentioned earlier.
2. Fill to mark with methylene chloride.
3. The concentration of analyte (C-15) and the recovery standard (C-17) in your sample is slightly low for accurate analysis on the GC, so we will concentrate it. This is a common approach in analytical chemistry. Pipet 10.00 mL of the 250-mL solution into a 10-mL Kuderna–Danish thimble.
4. Gently and carefully evaporate the CH_2Cl_2 to approximately 1.0 mL under a gentle stream of ultrahigh-grade He or N_2. To aid in the process, place the thimble in a warm (not boiling) water bath.
5. After the extraction solvent has reached approximately 1.0 mL, add 10 µL of the 3026-mg/L solution of C-13 (internal standard).
6. Wash down the walls of the thimble with a clean disposable pipet, and mix the solution completely.

7. Transfer the solution to a 1.5- to 2.0-mL autosample vial and seal with a Teflon-lined cap.

8. Analyze the sample on the GC (remember to sign in the logbook).

9. Conduct a linear least squares analysis using the spreadsheet from Chapter 2.

10. Calculate the concentration of C-15 in the original sample using the mass of soil weighed out, the GC results, your dilution factors, and the recoveries of your internal standards.

Results

Record all work in your laboratory notebook. Show all calculations. Write a two- to three-page report summarizing the results for the class. You should include a linear least squares analysis of your calibration data and a propagation of uncertainty analysis (see Chapter 2). Where does most of your uncertainty come from, your linear least squares analysis or your dilution/concentration steps? Your instructor will provide you with the known concentration of C-15 in your sample. Perform a Student's *t* test for your entire class results to see if your value is within the 95% confidence level.

ADVANCED STUDY ASSIGNMENT

This is an example of the Soxhlet Extraction calculation that you will be required to perform with your lab data. You extract a soil sample (50.57 g) for DDT in acetone. You want to check the recovery of DDT in your extraction procedure, so you add a recovery standard to the Soxhlet apparatus. After the Soxhlet extraction, you bring the final volume of solvent to 250 mL. Since the concentration of DDT in the solvent is too low to analyze by the GC, you concentrate 25.0 mL of the solvent (containing DDT) to 1.00 mL and add internal standard. The internal standard corrects for any injection errors and corrects the output from the GC automatically for any losses. You inject 1.00 μL of each standard and sample. The following data are collected:

Compound	Mass Added (pg)	Mass Recovered (pg)
Recovery standard	50.0	48.0
DDT	—	35.67
Internal standard	35.35	30.58

What is the concentration of DDT in your original 50.57-g sample?

DATA COLLECTION SHEET

DATA COLLECTION SHEET

DATA COLLECTION SHEET

DATA COLLECTION SHEET

DATA COLLECTION SHEET

17

DETERMINATION OF A CLAY–WATER DISTRIBUTION COEFFICIENT FOR COPPER

Purpose: To determine the distribution coefficient of a metal on a characterized soil

To learn to use a flame atomic absorption spectrometer

BACKGROUND

Perhaps the most important fate and transport parameter is the *distribution coefficient*, K_d, a measure of the adsorption phenomenon between the aqueous and solid phases and is fundamental to understanding the processes responsible for the distribution of pollutants in aquatic systems. (For its application to fate and transport modeling of groundwater, lakes, and riverine systems, refer to Chapters 24 through 27.) Mathematically, it can be represented as the ratio of the equilibrium pollutant concentration in the solid (sediment or soil) phase to the equilibrium pollutant concentration in the dissolved (aqueous) phase:

$$K_d = \frac{C_{\text{solid}}\,(\text{mg/kg})}{C_{\text{aqueous}}\,(\text{mg/L})} \tag{17-1}$$

The purpose of the distribution coefficient is to quantify which phase (solid or aqueous) the pollutant has a preference for and to determine the mass of pollutant present in each phase. The distribution coefficient is used in virtually every fate

Environmental Laboratory Exercises for Instrumental Analysis and Environmental Chemistry
By Frank M. Dunnivant
ISBN 0-471-48856-9   Copyright © 2004 John Wiley & Sons, Inc.

and transport model for the estimation of pollutant concentrations in aqueous systems. The aqueous-phase concentration is important because the free aqueous-phase concentration is usually the most toxic form of pollutants. Inorganic and organic colloids and suspended solids in natural waters will increase the apparent water-phase concentration, but pollutants adsorbed to these particles are usually not available for biological uptake. These particles can eventually settle out in quiescent regions of the natural water body or in estuaries and remove sufficient amounts of pollutant from the aquatic system.

Distribution coefficients are relatively easy to determine by allowing a pollutant–soil–water mixture of known composition to equilibrate, separating the mixture into solid and aqueous phases, and determining the pollutant concentration in each phase. This technique can be simplified by measuring (or knowing) the total mass of pollutant added to each sample (determined in a blank sample), measuring the pollutant in the dissolved phase after equilibration, and estimating the mass of pollutant on the solid phase by difference (total mass of pollutant in blank minus aqueous phase mass). The distribution coefficient is then calculated using equation (17-1).

The major problem with designing K_d experiments for the laboratory is the variability (and unpredictability) of results that are obtained given the variety of solid phases available, the nature of the pollutant used (ionic metals or hydrophobic organic compounds), and the experimental aqueous conditions used (pH values, ionic strengths, solids concentrations, and pollutant concentrations). Aqueous conditions are especially important when measuring K_d for ionic pollutants. Unless the lab instructor has conducted the experiment previously under the exact experimental conditions to be used in the lab, aqueous solutions may not contain sufficient pollutant in the aqueous phase to be measured, or all of the pollutant may be present in the aqueous phase. Given these experimental design problems, it is not surprising that this vital experimental parameter (K_d) is not typically taught in environmental chemistry lab courses but is usually covered in lecture material. In this chapter we present a procedure, using standardized materials and conditions, for the determination of a distribution coefficient for copper. The procedure is also environmentally friendly since no (or limited) hazardous waste is generated.

THEORY

The fate and transport of pollutants in aquatic systems and sorption phenomena of pollutants is often discussed in environmental chemistry lecture courses. How a pollutant interacts with its surroundings (solubility in water; sorption to soil and sediment particles) will greatly influence how it travels through the environment. *Sorption* is a generic term used to describe all processes in which a pollutant prefers a solid phase to the dissolved phase. Absorption and adsorption are used to describe this process for metals and polar (or ionizable) organics interacting with solid surfaces, while partitioning is used to describe this process for hydrophobic compounds interacting with natural organic matter. The key parameter describing absorption and adsorption is the distribution coefficient (K_d). The key parameter for describing partitioning is the *partition coefficient* (K_p). Both are ratios of the concentration of pollutant in or on the solid phase to the concentration of pollutant dissolved in the aqueous phase. The goal of this lab is to learn how to determine a distribution coefficient for a water–solid suspension containing Cu^{2+} ions. Normally, heavy metals are used in K_d determinations, but these generate hazardous waste and raise health concerns. Copper will be used in this laboratory exercise as a surrogate for heavy metals.

Adsorption of metals to clay surfaces is usually due to cationic exchange reactions resulting from a pH-dependent charge on the clay or from isomorphic substitutions. The pH-dependent charge is fairly self-explanatory and is present primarily on the broken edges of clays. Isomorphic substitution is a permanent charge on the clay resulting from Al or Si in the clay sheets being replaced by an element of lesser positive charge; thus, a net negative charge is present on the clay. This net negative charge is usually neutralized by common cations (i.e., Na^+, K^+, Ca^{2+}, Mg^{2+}, etc.) in solution, but some metals will preferentially exchange with these cations and be strongly held (adsorbed) to the clay surface. The extent of this adsorption is determined by the magnitude of the K_d.

ACKNOWLEDGMENTS

I would like to thank Jason Kettel (Whitman College, Class of 2001) for designing and collecting data for this experiment. I am also indebted to the Whitman environmental chemistry class of spring 2002 for refining the procedures of this experiment.

REFERENCES

O'Connor, D. J. and J. P. Connolly, *Water Res.*, **14**, 1517–1523 (1980).

Stumm, W. and J. J. Morgan, *Aquatic Chemistry: An Introduction Emphasizing Chemical Equilibria in Natural Waters*, 2nd ed., Wiley, New York, 1981.

IN THE LABORATORY

There are a number of ways to conduct this laboratory exercise; your instructor will decide which is best for your class. One option is to divide the class into three groups, one group for the determination of K_d as a function of Cu concentration, one group for the determination of K_d as a function of ionic strength, and one group for the determination of K_d as a function of suspended solids. Another option is to have the entire class determine K_d as a function of Cu concentration; in this case, the instructor may provide you with the results for the other experiments. Regardless of the approach being used, you must come to lab with a good understanding of K_d and how the experiments are designed.

Safety Precautions

- As in all laboratory exercises, safety glasses must be worn at all times.
- Avoid skin and eye contact with NaOH, HCl, and HNO_3 solutions. If contact occurs, rinse your hands and/or flush your eyes for several minutes. Seek immediate medical advice for eye contact.

Chemicals and Solutions

- Sorbents
  - *Ca-montmorillinite* (obtained from the Clay Minerals Society, Source Clays Repository. Product STx-1. The origin of the clay, chemical composition, cation exchange capacity, and surface area are given at `http://cms.land.gov` and in the supplemental material of this article. Ordering information is also available at this Web site. Mass requirements are about 0.100 ± 0.001 g per vial.)
  - *K-kaolinite* (obtained from the Clay Minerals Society, Source Clays Repository. Product KGa-1b. The origin of the clay, chemical composition, cation exchange capacity, and surface area are given at `http://cms.land.gov` and in the supplemental material of this article. Ordering information is also available at this Web site. Mass requirements are about 0.100 ± 0.001 g per vial.)
- *Stock copper solution.* Dissolve 0.268 g of $CuCl_2 \cdot 2H_2O$ (GMW 170.34) in 100.0 mL of deionized water (volumetric flask) to make a 1000-mg/L solution. Make a $1 : 100$ dilution of this solution to obtain a 10.0-mg/L solution of Cu^{2+}.
- *Calcium nitrate solution.* Prepare two 100-mL portions of $2.00\,M$ $Ca(NO_3)_2 \cdot 4H_2O$.
- *$Ca(NO_3)_2$ ionic strength adjustor.* $0.8469\,M$: Dissolve 11.808 g of $Ca(NO_3)_2 \cdot 4H_2O$ (GMW 236.16 g/mol) in 100.0 mL of deionized water to

make a $0.500\,M$ $Ca(NO_3)_2 \cdot 4H_2O$ solution. Make $10:100$ and $1:100$ dilutions of this solution to obtain $0.08469\,M$ and $0.008469\,M$ solutions, respectively.

- $1\,M$ HCl and $1\,M$ NaOH for adjusting pH.
- 1% nitric acid.
- Run blanks of each Cu solution in equilibrium vials.
- 1000 mg/L Cu standard in 5% HNO_3.

Equipment and Glassware

- Cu flame atomic absorption spectroscopy (FAAS) lamp
- FAAS unit
- New 50-mL plastic, sealable vials (24 vials per solid evaluated are needed) (blue max. disposable centrifuge tubes, polystyrene, conical bottom, sterile: Falcon, VWR Scientific Products Number 21008-939)
- 100- or 50-mL graduated cylinders
- Plastic filter holders and filters (polycarbonate filter holder, 25-mm filter,12 per pack; VWR Scientific Products Number 22001-800)
- 25-mL plastic syringes
- 25-mm Gelman-type A/E glass fiber filter (glass fiber filters, type A/E; Pall Gelman, VWR Scientific Products Number 28150-178)
- 0.2-μm membrane filter or similar brand (Spartan-13, Agilent Technologies, HP-5061-3366)
- Plastic beakers for holding filtered samples
- Test tube rotator (Glas-Col mini-rotator, 120 V, VWR Scientific Products Number 33725-042; test tube rockers will probably work just as well)

PROCEDURE

Week 1

Prerinse all plastic vials and caps with deionized water.

Team 1: K_d as a Function of Total Suspended Solids (TSS) and Clay Type

1. The mineral phases to be used as your adsorbent are kaolinite (KGa-1b) and Ca-montmorillinite (STx-1). Thus, you will have two sets of vials, or two experiments, one with each absorbent.

2. *Preparation of stock Cu solution* (from $CuCl_2 \cdot 2H_2O$). Make a 1000-mg/L solution by adding 2.683 g of $CuCl_2 \cdot 2H_2O$ to a 1-L volumetric flask and filling to the mark with deionized water. (Do not add acid yet.)

3. *Preparation of solutions for making suspensions.* The goal is to prepare solutions in which the ionic strength and pH are as close to identical as possible. There are probably several ways to do this, but we will use the following approach:

    ○ *$Ca(NO_3)_2 \cdot 4H_2O$ stock solution.* Transfer 29.54 g of $Ca(NO_3)_2 \cdot 4H_2O$ (GMW is 236.16 g/mol) to a 250-mL volumetric flask and fill to the mark. This will yield a $0.500\,M$ solution.

4. Your goal is to measure K_d as a function of TSS and mineral phase. Prepare two vials for each TSS concentration of each clay type. You will use four different TSS concentrations of each clay: 500 mg/L, 1000 mg/L, 5000 mg/L, and 10,000 mg/L. You will be using a total volume in each sample vial of 40.0 mL. Weigh 0.020 g (for the 500-mg/L vials), 0.0400 g (for the 1000-mg/L vials), 0.200 g (for the 5000-mg/L vials), and 0.400 g (for the 10,000-mg/L vials). Be as close as you can to these weights, and record your significant figures to four decimal places. All vials in this experiment will use a copper concentration of 5.00 mg/L. You will also need to have two blanks containing ionic strength adjustor, Cu, and water (see step 5), but no mineral phase. Label each with masking tape and a number (e.g., "T1-1" represents "team 1 vial 1," "T1-B1," "team 1, blank 1"). In all, you will have at least two blanks (no mineral phase) and two vials for each TSS of each mineral phase.

5. Prepare the following solution in a 100-mL (or better yet, 50-mL) graduated cylinder:

    ○ 2.00 mL of 0.50 M $Ca(NO_3)_2 \cdot 4H_2O$ stock solution.

    ○ Add the appropriate amount of Cu solution (for this experiment, consult the 5.00-ppm row in Table 17-1).

    ○ Fill to 40.0 mL with deionized water.

6. Add the solution to each vial prepared in step 5, cap, and mix well. The pH should be between 5.0 and 5.5 for the kaolinite and between 6.5 and 7.0 for the montmorillinite. Adjust as needed with $1M$ HCl or NaOH.

TABLE 17-1. Cu Solution Table for Team 1

Desired Cu Solution Concentration in a Vial (ppm)	Addition Volume (mL) of the Cu Solution to the Right to Yield the Desired Cu Concentration to the Left	Standard Cu Solution[a] (mg/L)
50.0	2.00	1000.
25.0	1.00	1000.
10.0	4.00	100.
5.00	**2.00**	**100.**
1.00	4.00	10.0
0.500	2.00	10.0

[a]To prepare a 1000-ppm Cu^{2+} solution, add 2.683 g of $CuCl_2 \cdot 2H_2O$ to a 1000–mL volumetric flask and fill to the mark. To prepare the 100-ppm Cu solution, make a 10 : 100 dilution of the 1000-ppm solution. To prepare the 10-ppm Cu solution, make a 1 : 100 dilution of the 1000-ppm solution.

7. Again, be sure to prepare at least two blanks for each Cu concentration (containing everything, including Cu standard, but no solid phase). These will be necessary to determine if any Cu adsorbs to the container walls.

8. Place the vials on the mixer for at least three days.

Team 2: K_d as a Function of Cu Concentration (Kaolinite)

1. The mineral phase to be used as your adsorbent is kaolinite (KGa-1b).

2. *Preparation of stock Cu solution* (from $CuCl_2 \cdot 2H_2O$). Make a 1000-mg/L solution by adding 2.683 g of $CuCl_2 \cdot 2H_2O$ to a 1-L volumetric flask and filling to the mark with deionized water. (Do not add acid yet.)

3. *Preparation of solutions for making suspensions.* The goal is to prepare solutions where the mass of solid phase, the ionic strength, and the pH are as close to identical as possible. There are probably several ways to do this, but we will use the following approach:

   o $Ca(NO_3)_2 \cdot 4H_2O$ stock solution. Transfer 29.54 g of $Ca(NO_3)_2 \cdot 4H_2O$ (GMW is 236.16 g/mol) to a 250-mL volumetric flask and fill to the mark. This will yield a 0.500 M solution.

   Your goal is to measure the K_d as a function of Cu concentration for a kaolinite clay. Prepare two vials for each Cu concentration. You will use a TSS concentration of 5000 mg/L. You will be using a total volume in each sample vial of 40.0 mL. Weigh 0.200 g (for the 5000 mg/L TSS) into each vial (except your blank vials). Be as close as you can to this mass, and record your significant figures to four decimal places. You will also need to have two blanks for each Cu concentration. These blank vials will contain ionic strength adjustor, Cu, and water (see step 4), but no mineral phase. Label each with masking tape and a number (e.g., "T2-1" represents "team 2, vial 1"; "T2-B1," "team 2, blank 1").

TABLE 17-2. Cu Solution Table for Team 2

Desired Cu Solution Concentration in a Vial (ppm)	Addition Volume (mL) of the Cu Solution to the Right to Yield the Desired Cu Concentration to the Left	Standard Cu Solution[a] (mg/L)
50.0	2.00	1000.
25.0	1.00	1000.
10.0	4.00	100.
5.00	2.00	100.
1.00	4.00	10.0
0.500	2.00	10.0

[a] To prepare a 1000-ppm Cu^{2+} solution, add 2.683 g of $CuCl_2 \cdot 2H_2O$ to a 1000-mL volumetric flask and fill to the mark. To prepare the 100-ppm Cu solution, make a 10 : 100 dilution of the 1000-ppm solution. To prepare the 10-ppm Cu solution, make a 1 : 100 dilution of the 1000-ppm solution.

4. Prepare the following solutions to fill the sediment-containing vials and blanks in a 100-mL (or better yet, 50-mL) graduated cylinder, using Cu^{2+} solutions of 1000 ppm, 100 ppm, and 10 ppm, made as described in Table 17-2.
   ○ 2.00 mL of 0.50 M $Ca(NO_3)_2 \cdot 4H_2O$ stock solution.
   ○ Add the appropriate amount of Cu solution for each concentration (Table 17-2).
   ○ Fill to 40.0 mL with deionized water.

5. Add each solution to the appropriate vials, cap, and mix well.

6. Again, be sure to prepare two blanks for each Cu concentration (containing everything, including Cu standard, but no solid phase). These will be necessary to determine if any Cu adsorbs to the container walls.

7. Place the vials on the mixer for at least three days.

Team 3: K_d as a Function of Cu Concentration (Montmorillinite (STx-1)

1. The mineral phase to be used as your adsorbent is montmorillinite (STx-1).

2. *Preparation of stock Cu solution* (from $CuCl_2 \cdot 2H_2O$). Make a 1000-mg/L solution by adding 2.683 g of $CuCl_2 \cdot 2H_2O$ to a 1-L volumetric flask and filling to the mark with deionized water. (Do not add acid yet.)

3. *Preparation of solutions for making suspensions.* The goal is to prepare solutions where the mass of solid phase, ionic strength, and pH are as close to identical as possible. There are probably several ways that we can do this, but we will use the following approach.
   ○ *$Ca(NO_3)_2 \cdot 4H_2O$ stock solution.* Transfer 29.54 g of $Ca(NO_3)_2 \cdot 4H_2O$ (GMW is 236.16 g/mol) to a 250-mL volumetric flask and fill to the mark. This will yield a 0.500 M solution.

TABLE 17-3. Cu Solution Table for Team 3

Desired Cu Solution Concentration in a Vial (ppm)	Addition Volume (mL) of the Cu Solution to the Right to Yield the Desired Cu Concentration to the Left	Standard Cu Solution[a] (mg/L)
50.0	2.00	1000.
25.0	1.00	1000.
10.0	4.00	100.
5.00	2.00	100.
1.00	4.00	10.0
0.500	2.00	10.0

[a] To prepare a 1000-ppm Cu^{2+} solution, add 2.683 g of $CuCl_2 \cdot 2H_2O$ to a 1000-mL volumetric flask and fill to the mark. To prepare the 100-ppm Cu solution, make a 10 : 100 dilution of the 1000-ppm solution. To prepare the 10-ppm Cu solution, make a 1 : 100 dilution of the 1000-ppm solution.

Your goal is to measure K_d as a function of Cu concentration for a montmorillinite clay. Prepare two vials for each Cu concentration. You will use a TSS concentration of 5000 mg/L. You will be using a total volume in each sample vial of 40.0 mL. Weigh 0.200 g (for the 5000-mg/L TSS vials) in each vial (except your blank vials). Be as close as you can to this mass, and record your significant figures to four decimal places. You will also need to have two blanks for each Cu concentration. These blank vials will contain ionic strength adjustor, Cu, and water (see step 4), but no mineral phase. Label each with masking tape and a number (e.g., "T3-1" represents "team 3, vial 1"; "T3-B1" "team 3, blank 1").

4. Prepare the following solutions in a 100-mL (or better yet, 50-mL) graduated cylinder:

   ○ 2 mL of 0.50 M $Ca(NO_3)_2 \cdot 4H_2O$ stock solution.

   ○ Add the appropriate amount of Cu solution from Table 17-3.

   ○ Fill to 40.0 mL with deionized water.

5. Add each solution to the appropriate vial, cap, and mix well.

6. Again, be sure to prepare two blanks for each Cu concentration (containing everything, including Cu standard, but no solid phase). These will be necessary to determine if any Cu adsorbs to the container walls.

7. Place the vials on the mixer for at least three days.

Team 4: K_d as a Function of Ionic Strength (I) and Mineral Phase

1. The mineral phases to be used as your adsorbent are kaolinite and montmorillinite.

2. *Preparation of stock Cu solution* (from $CuCl_2 \cdot 2H_2O$). Make a 1000-mg/L solution by adding 2.683 g of $CuCl_2 \cdot 2H_2O$ to a 1-L volumetric flask and filling to the mark with deionized water. (Do not add acid yet.)

TABLE 17-4. Cu Solution Table for Team 4

Desired Cu Solution Concentration in a Vial (ppm)	Addition Volume (mL) of the Cu Solution to the Right to Yield the Desired Cu Concentraction to the Left	Standard Cu Solution[a] (mg/L)
50.0	2.00	1000.
25.0	1.00	1000.
10.0	4.00	100.
5.00	2.00	100.
1.00	4.00	10.0
0.500	2.00	10.0

[a] To prepare a 1000-ppm Cu^{2+} solution, add 2.683 g of $CuCl_2 \cdot 2H_2O$ to a 1000-mL volumetric flask and fill to the mark. To prepare the 100-ppm Cu solution make a 10 : 100 dilution of the 1000-ppm solution. To prepare the 10-ppm Cu solution, make a 1 : 100 dilution of the 1000-ppm solution.

3. *Preparation of solutions for making suspensions.* The goal of this is to prepare solutions where the mass of solid phase and that of Cu concentration are identical while the ionic strength changes systematically. There are probably several ways to do this, but we will use the following approach:

   ○ *Ca(NO₃)₂·4H₂O stock solution.* Transfer 29.54 g of $Ca(NO_3)_2 \cdot 4H_2O$ (GMW is 236.16 g/mol) to a 250-mL volumetric flask and fill to the mark. This will yield a 0.500 M solution.

   Your goal is to measure K_d as a function of ionic strength (I) for a kaolinite and montmorillinite clay. Prepare two vials for each ionic strength and clay type. You will use a TSS concentration of 5000 mg/L and a total volume in each sample vial of 40.0 mL. Weigh 0.200 g (for 5000 mg/L TSS) into each vial (except your blank vials). Be as close as you can to this weight, and record your significant figures to four decimal places. You will also need to have two blanks. These blank vials will contain ionic strength adjustor, Cu, and water (see step 4), but no mineral phase. Label each with masking tape and a number (e.g., "T4-1" represents "team 4, vial 1"; "T4-B1," "team 4, blank 1").

4. Prepare the following solution in a 100-mL (or better yet, 50-mL) graduated cylinder:

   ○ Use the appropriate amount of Cu solution (for you, this will be 5.00 ppm in Table 17-4).

   ○ Add $Ca(NO_3)_2 \cdot 4H_2O$ stock solution. (Determine the appropriate amount from Table 17-5. You will need to have the appropriate dilutions shown in the second column.)

   ○ Fill to 40.0 mL with deionized water.

5. Add each solution to the appropriate vials, cap, and mix well.

TABLE 17-5. Table for Determining the Ionic Strength of the Solution for Team 4

Addition (mL)	of a Molar $Ca(NO_3)_2$ Solution (mol/L) to 100 mL	to Obtain a Final $Ca(NO_3)_2$ (mg/L) Concentration of:	Final Ionic Strength (mg/L)
2.00	0.008469	100	9,600
1.00	0.08469	500	10,900
2.00	0.08469	1,000	12,400
1.00	0.8469	5,000	24,400
2.00	0.8469	10,000	39,400
3.00	0.8469	15,000	54,400
4.00	0.8469	20,000	69,400

6. Again, be sure to prepare two blanks (containing everything, including Cu standard, but no solid phase). These will be necessary to determine if any Cu adsorbs to the container wall.
7. Place the vials on the mixer for at least three days.

Week 2

There will be several demonstrations at the beginning of lab to illustrate use of the filter apparatus and mixing system.

1. Turn on the AAS to warm up the lamp.
2. Prepare calibration standards at concentrations of 0.100, 0.500, 1.00, 5.00, 10.0, 25.0, and 50.0 ppm Cu^{2+}. Prepare these in 1% HCl.
3. Filter the solutions that you prepared last week. First, filter them through the Gelman-type A/E glass-fiber filter, then through a 0.2-μm HPLC nylon filter with a syringe. Filter both the blanks and the actual samples.
4. Analyze the samples using AAS as demonstrated.
5. Turn in your data in tabular form and as a graph.

Waste Disposal

After neutralization, all solutions can be disposed of down the drain with water.

ASSIGNMENT

For your lab report, compile all of the data for each solid, estimate K_d for each solid phase, and write a short answer to each of the following issues.

1. Contrast the differences in K_d between the solid phases.
2. Contrast the results for the variation of TSS.
3. Contrast the results for the variation of ionic strength.
4. Explain why the dilution water contained $Ca(NO_3)_2$.

ADVANCED STUDY ASSIGNMENT

1. Prepare a list of things to do when you arrive in the laboratory.

2. Prepare a dilution table showing how you will make your calibration standards for the flame atomic absorption spectroscopy unit.

3. Research the clay mineralogy and structure of kaolinite and montmorillinite. Turn in chemical formulas and a figure of the structures. Show how montmorillinite can undergo isomorphic substitution.

4. Draw and label the major components of a flame atomic absorption spectrometer. Describe each major component in two to three sentences.

DATA COLLECTION SHEET

DATA COLLECTION SHEET

DATA COLLECTION SHEET

DATA COLLECTION SHEET

DATA COLLECTION SHEET

PART 6

WET EXPERIMENTS

18

DETERMINATION OF DISSOLVED OXYGEN IN WATER USING THE WINKLER METHOD

Purpose: To determine the dissolved oxygen concentration in a water sample

To learn the chemical reactions involved in the Winkler dissolved oxygen method

BACKGROUND

It is a common perception that all life is dependent on the presence of oxygen, either in the atmosphere or in the water. However, this is anything but true. The first life-forms to evolve on Earth are thought to have been anaerobic, requiring an oxygen-free environment to grow. In fact, free oxygen is toxic to anaerobic organisms' biochemical machinery. Oxygen was actually a waste product from these organisms and through the emission of oxygen over hundreds of millions of years enabled the evolution of aerobic organisms. Even today there are many types of respiration (and organisms) that do not require the presence of oxygen as their terminal electron acceptor (TCE). Every life-form needs a terminal electron acceptor to accept the excess electrons from their reduced food sources. For example, look at how we oxidize glucose with atmospheric oxygen to yield energy (the first reaction in Table 18-1). Electrons on glucose arc removed and added to diatomic oxygen, and in this process oxygen is reduced from an oxidation state of zero to -2 while carbon is oxidized to $+4$. The net result is a generation of 2863 kJ of energy per mole of glucose oxidized, a higher energy yield than that

Environmental Laboratory Exercises for Instrumental Analysis and Environmental Chemistry
By Frank M. Dunnivant
ISBN 0-471-48856-9   Copyright © 2004 John Wiley & Sons, Inc.

TABLE 18-1. Energy Evolved by Different Terminal Electron Acceptors

Balanced Reactions of Substrate and Each TCE	Net $\Delta G^{\circ}_{H_2O}$ (kJ/mol)
$6\,O_2 + C_6H_{12}O_6 \Rightarrow 6\,H_2O + 6\,CO_2$	-2863
$3\,NO_3^- + C_6H_{12}O_6 + 6\,H^+ \Rightarrow 3\,NH_4^+ + 6\,CO_2 + 3\,H_2O$	-1817
$3\,SO_4^{2-} + C_6H_{12}O_6 + 3\,H^+ \Rightarrow 3\,HS^- + 6\,H_2O + 6\,CO_2$	-473
$3\,CO_2 + C_6H_{12}O_6 \Rightarrow 3\,CH_4 + 6\,CO_2$	-420

Source: Calculated from data in Schwarzenbach et al. (1993).

achieved with more primitive TCEs. This is one reason why organisms that use oxygen as their TCE outcompete other life-forms. The first life-forms yielded only small amounts of energy from their oxidation of food substrates. In doing so, some of these organisms (not shown) produced oxygen and created our oxygen-abundant atmosphere, which allowed aerobic life-forms to evolve.

The reactions shown in Table 18-1 represent the transitions of TCEs in water as the environment changes from aerobic to anaerobic. First, oxygen is used since it produces the most energy per mole of glucose oxidized. This is followed by nitrate, then sulfate, and finally, carbon dioxide. Other possible TCEs include metal ions such as Fe^{3+}.

There are basically two living environments on Earth, those with and those without free diatomic oxygen. Table 18-2 shows average ranges of reduction potentials (E_H) and pH values for different sources of water. The reduction potential (E_H) reflects the presence or absence of dissolved oxygen (DO). Oxygenated waters have more positive E_H values; waters with low oxygen and anaerobic waters have negative E_H values.

Oxygen is considered poorly soluble in water. It is interesting to note that air-breathing organisms have around 19% oxygen (262 mg/L from $PV = nRT$) to consume, whereas organisms respiring in water have only a maximum of about

TABLE 18-2. Average E_H and pH Values for Water Samples

Water Source	E_H (V)	pH
Mine waters	0.60	4.0
Rainwater	0.55	5.0
Stream waters	0.40	6.1
Normal ocean waters	0.30	9.0
Aerated saline water residues	0.20	10
Groundwaters	-0.15	9.0
Bog waters	0.0	6.0
Waterlogged soils	-0.10	3.0
Organic-rich saline waters	-0.40	9.0

Source: Garrels and Christ (1990).

0.15% oxygen (14.6 mg/L). As temperature increases or as salt content increases, the DO concentration decreases. The range of DO concentrations in water under normal conditions is shown in Table 18-3. Note that the range of DO in pure water (no salt content) is from 7.6 mg/L at 30°C to 14.6 mg/L at 0°C. Although this may seem like a narrow range, many organisms have become specialized so that they can live in only a small portion of this range. Important examples are mountain trout and several species of invertebrate insect larva, which require very cold waters with the highest concentrations of dissolved oxygen.

TABLE 18-3. Solubility of Dissolved Oxygen for Water in Contact with a Dry Atmosphere[a]

Temperature (°C)	Chloride Concentration (mg/L)				
	0	5,000	10,000	15,000	20,000
0	14.6	13.8	13.0	12.1	11.3
1	14.2	13.4	12.6	11.8	11.0
2	13.8	13.1	12.3	11.5	10.8
3	13.5	12.7	12.0	11.2	10.5
4	13.1	12.4	11.7	11.0	10.3
5	12.8	12.1	11.4	10.7	10.0
6	12.5	11.8	11.1	10.5	9.8
7	12.2	11.5	10.9	10.2	9.6
8	11.9	11.2	10.6	10.0	9.4
9	11.6	11.0	10.4	9.8	9.2
10	11.3	10.7	10.1	9.6	9.0
11	11.1	10.5	9.9	9.4	8.8
12	10.8	10.3	9.7	9.2	8.6
13	10.6	10.1	9.5	9.0	8.5
14	10.4	9.9	9.3	8.8	8.3
15	10.2	9.7	9.1	8.6	8.1
16	10.0	9.5	9.0	8.5	8.0
17	9.7	9.3	8.8	8.3	7.8
18	9.5	9.1	8.6	8.2	7.7
19	9.4	8.9	8.5	8.0	7.6
20	9.2	8.7	8.3	7.9	7.4
21	9.0	8.6	8.1	7.7	7.3
22	8.8	8.4	8.0	7.6	7.1
23	8.7	8.3	7.9	7.4	7.0
24	8.5	8.1	7.7	7.3	6.9
25	8.4	8.0	7.6	7.2	6.7
26	8.2	7.8	7.4	7.0	6.6
27	8.1	7.7	7.3	6.9	6.5
28	7.9	7.5	7.1	6.8	6.4
29	7.8	7.4	7.0	6.6	6.3
30	7.6	7.3	6.9	6.5	6.1

Source: After Wipple and Wipple (1911).
[a](Water at 1.0 atm containing 20.9% oxygen) (increasing the salt content of water decreases the solubility of any dissolved gas).

THEORY

Two methods are commonly used to determine the concentration of dissolved oxygen in water samples: the Winkler or iodometric method and the membrane electrode technique. Details on each of these methods can be found in standard methods of the American Water Works Association (AWWA, 1998) and in Sawyer and McCarty (1978). The iodometric method, discussed first, is the focus of this laboratory procedure. A more recent development is the use of the Ru(bipy)$_3$ optical sensor for O$_2$. Information regarding the latter sensor can be found on the Internet.

The iodometric method, the more accurate of the two methods, determines the dissolved oxygen concentration through a series of oxidation–reduction reactions. First, Mn^{2+} (as MnSO$_4$) is added to a 250- or 300-mL sample. Next, the alkali–iodide reagent (KI in NaOH) is added. Under these caustic conditions, if oxygen is present in the water sample, the Mn^{2+} will be oxidized to Mn^{4+}, which precipitates as a brown hydrated oxide. This reaction is relatively slow and the solution must be shaken several times to complete the reaction. This reaction can be represented by the following equations:

$$2\,Mn^{2+} + 4\,OH^- + O_2 \rightarrow 2\,MnO_2(s) + 2\,H_2O$$

or

$$2\,Mn(OH)_2 + O_2 \rightarrow 2\,MnO_2(s) + 2\,H_2O$$

After the MnO$_2$ precipitate settles to the bottom of the flask, sulfuric acid is added to make the solution acidic. Under these low-pH conditions, the MnO$_2$ oxidizes the iodide (I$^-$) to free iodine (I$_2$) through the reaction

$$MnO_2 + 2\,I^- + 4\,H^+ \rightarrow Mn^{2+} + I_2 + 2\,H_2O$$

Now the sample is ready for titration with standardized sodium thiosulfate (Na$_2$S$_2$O$_3$·5H$_2$O). In this reaction, thiosulfate ion is added quantitatively to convert the I$_2$ back to I$^-$. The amount of I$_2$ present at this stage in the procedure is directly related to the amount of O$_2$ present in the original sample. The reaction can be represented by

$$2\,S_2O_3^{2-} + I_2 \rightarrow S_4O_6^{2-} + 2\,I^-$$

The titration is complete when all of the I$_2$ has been converted to I$^-$. The endpoint of this titration can be determined through potentiometry or by using calorimetric indicators. The most common indicator is starch, which turns from deep blue to clear.

The DO concentration can be determined using the following equation, which also reflects the series of redox reactions in the equations given above:

$$mg\ O_2/L = \frac{\left[(L\ S_2O_3^{2-})(M\ S_2O_3^{2-})\right]\left(\dfrac{I_2}{2\,S_2O_3^{2-}}\right)\left(\dfrac{MnO_2}{I_2}\right)\left(\dfrac{O_2}{2\,MnO_2}\right)\left(\dfrac{32\,g}{mol\ O_2}\right)\left(\dfrac{1000\,mg}{g\ O_2}\right)}{L\ sample}$$

$$(18\text{-}1)$$

Several modifications of the Winkler method have been developed to overcome interferences. The azide modification, the most common modification, effectively removes interference from nitrite, which is commonly present in water samples from biologically treated wastewater effluents and incubated biochemical oxygen demand samples. Nitrite interferes by converting I^- to I_2, thus overestimating the dissolved oxygen in the sample. This is illustrated in the following equations:

$$2\,NO_2^- + 2\,I^- + 4\,H^+ \rightarrow I_2 + N_2O_2 + 2\,H_2O$$
$$N_2O_2 + \tfrac{1}{2}O_2 + H_2O \rightarrow 2\,NO_2^- + 2\,H^+$$

Note that N_2O_2 is oxidized by oxygen, which enters the sample during the titration procedure and is converted to NO_2^- again, establishing a cyclic reaction that can lead to erroneously high results. This final result yields oxygen concentrations that are far in excess of the amounts that would normally be expected.

Nitrite interference can easily be overcome through the addition of sodium azide (NaN_3). Azide is generally added with the alkali–KI reagent, and when sulfuric acid is added, the following reactions result in the removal of NO_2^-:

$$NaN_3 + H^+ \rightarrow HN_3 + Na^+$$
$$HN_3 + NO_2^- + H^+ \rightarrow N_2 + N_2O + H_2O$$

Other methods can also be used to remove ferrous iron (the permanganate modification), ferric iron (the potassium fluoride modification), and suspended solids (the alum flocculation modification). We will be using only the azide modification in this laboratory experiment.

The electrode method offers several advantages over the titration method, including speed, elimination or minimization of interferences, field compatibility, continuous monitoring, and in situ measurement. However, some loss in accuracy results. Modern electrodes rely on a selectively permeable membrane that allows dissolved oxygen to enter the measurement cell, thus eliminating most interferences. A detailed description of the operation of this electrode can be found in Sawyer and McCarty (1978). The calibration and measurement is relatively simple, and a direct readout of the oxygen concentration (in mg/L) is given.

REFERENCES

American Water Works Association, *Standard Methods for the Examination of Water and Wastewater*, 20th ed., AWWA, Denver, CO, 1998.

Garrels, R. M. and C. L. Christ, *Solutions, Minerals, and Equilibria*, Harper & Row, New York, 1990.

Sawyer, C. N., and P. L. McCarty. *Chemistry for Environmental Engineering*, 3rd ed., McGraw-Hill, New York, 1978.

Schwarzenbach, R. P., P. M. Gschwend, and D. M. Imboden, *Environmental Organic Chemistry*, Wiley, New York, 1993, Table 12.16.

Wipple, G. C. and M. C. Wipple, *J. Am. Chem. Soc.*, **33**, 362 (1911).

IN THE LABORATORY

You will be given one or more samples by your instructor for titration using the Winkler method. For this laboratory exercise you do not have to be concerned with preservation of the sample or sample-handling practices, but in the real world many precautions need to be taken. Most important is the preservation of field samples that need to be analyzed in the laboratory. The easiest way to avoid this is to use a field meter to determine the concentration of DO. This method is quick and relatively reliable. However, DO meters are expensive, and some monitoring programs may require you to use the Winkler titration method because of its greater accuracy.

Two approaches are used to preserve samples for later DO determination. First, you can "fix" your samples using the procedures described below and then perform the titration when the samples are brought to the laboratory. Samples should be stored in the dark and on ice until titration. This preservation technique will allow you to delay the titration for up to 6 hours. However, this procedure may give low results for samples with a high iodine demand. In this case it is advisable to use the second option, which is to add 0.7 mL of concentrated sulfuric acid and 0.02 g of sodium azide. When this approach is used, it is necessary to add 3 mL of alkali–iodide reagent (below) rather than the usual 2 mL. In addition, avoid any sample treatment or handling that will alter the concentration of DO, including increases in temperature and the presence of atmospheric headspace in your sample container.

You will titrate your samples using the procedures described below. As in all titration experiments, you should do a quick titration to determine the approximate volume of titrant needed. Follow this first titration with at least three careful titrations. Average your values for each sample.

Safety Precautions

- As in all laboratory exercises, safety glasses must be worn at all times.
- Avoid skin and eye contact with caustic and acidic solutions. If contact occurs, rinse your hands and/or flush your eyes for several minutes. Seek immediate medical advice for eye contact.
- Use concentrated acids in the fume hood and avoid breathing their vapors.
- Sodium azide is a toxin and should be treated as such.

Chemicals and Solutions

- Manganese sulfate: Dissolve 480 g of $MnSO_4 \cdot 4H_2O$, 400 g of $MnSO_4 \cdot 2H_2O$, or 364 g of $MnSO_4 \cdot H_2O$ in about 800 mL of deionized water. Filter the solution and dilute to 1.0 L.

- Alkali–iodide–azide reagent. Dissolve 500 g of NaOH (or 700 g of KOH) and 135 g of NaI (or 150 g of KI) in deionized water and dilute to 1.0 L. Add 10 g of NaN_3 dissolved in 40 mL of deionized water.

- Concentrated H_2SO_4. (1.0 mL of this solution is equivalent to approximately 3 mL of alkali–iodide–azide solution.)

- Starch solution. Dissolve 2 g of laboratory-grade soluble starch and 0.2 g of salicylic acid (as a preservative) in 100 mL of hot distilled water. Allow to cool before use.

- Sodium thiosulfate titrant, 0.0250 M. Dissolve 6.205 g of $Na_2S_2O_3 \cdot 5H_2O$ in deionized water. Add 1.5 mL of 6 M NaOH or 0.4 g of solid NaOH and dilute to 1.0 L. Standardize with biiodate solution.

- Standard potassium biiodate solution, 0.00210 M. Dissolve 0.8124 g of $KH(IO_3)_2$ in deionized water and dilute to 1.000 L.

Glassware

For each student group:

- Four Erlenmeyer flasks
- 25-mL buret
- 20.00-mL pipet
- Pasteur pipets
- Three 1.00-mL pipets (at least one of these should be a wide-bore pipet for the viscous azide reagent)

PROCEDURE

Standardization of Sodium Thiosulfate Titrant

Note: The thiosulfate titrant may already have been standardized by your instructor. If so, skip to step 5.

1. Dissolve approximately 2 g of KI (free of iodate) in an Erlenmeyer flask containing 100 to 150 mL of deionized water.
2. Add 1 mL of 6 M H_2SO_4 or a few drops of concentrated H_2SO_4 and pipet 20.00 mL of standard biiodate solution into the flask. Recall from the reactions given in the theory section that I_2 will be formed from the reaction when any DO is present.
3. Titrate the liberated I_2 with thiosulfate titrant until a pale straw (yellow) color is reached. Add a few drops of starch indicator, which will result in a blue color, and continue the titration to the endpoint, which is clear.
4. If all solutions were made properly, 20.00 mL of the biiodate solution will require 20.00 mL of thiosulfate titrant. If this result is not achieved, calculate the exact molar concentration of your titrant.

Titration of Water Samples

5. To a 250- or 300-mL sample bottle, add 1 mL of $MnSO_4$ solution, followed by 1 mL of alkali–iodide–azide reagent. If your pipets are dipped into the sample (as they should be), rinse them before returning them to the reagent bottles. If the solution turns white, no DO is present.
6. Stopper the sample bottles in a manner to exclude air bubbles and mix by inverting the bottle rapidly a few times. When the precipitate has settled to half the bottle volume, repeat the mixing and allow the precipitate to resettle.
7. Add 1.0 mL of concentrated H_2SO_4.
8. Restopper and mix by inverting the bottle rapidly and dissolve the precipitate. You may open the bottle and pour the sample at this point since the DO and reagents have been "fixed" and will not react further.
9. Titrate 200 mL of the sample with your standardized thiosulfate solution. Again, first titrate to a pale straw color, add starch indicator, and titrate to a clear endpoint.
10. Repeat the titration for two more samples and average your results.

Waste Disposal

After neutralization, all solutions can be disposed of down the drain with water.

ASSIGNMENT

1. Create a flowchart showing all of the oxidation–reduction reactions involved in the Winkler titration method. Explain each reaction.
2. Calculate an average and a standard deviation for each sample.

ADVANCED STUDY ASSIGNMENT

1. Why is dissolved oxygen important in aquatic environments?
2. What range of DO values would you expect for natural water samples?
3. What unit of measure is DO expressed in?
4. Table 18-3 is given for a dry atmosphere. How would the values given in this table change if you had an atmosphere with high humidity?
5. List two methods that can be used to determine DO.
6. Review the reagents used to fix the oxygen. Which reagents are critical (must be added in a quantitative manner), and which are not critical?
7. What is the purpose of the sodium azide modification to the Winkler titration procedure?
8. What is the color change for the starch indicator?
9. Briefly outline a procedure for titrating a water sample for DO. (List the major steps.)
10. Using your knowledge of stoichiometry, show how 1.00 mL of 0.025 M thiosulfate solution is equal to 1.00 mg/L DO for your 200 mL sample.
11. You titrate 200 mL of a sample with 0.0250 M thiosulfate and the titration takes 8.65 mL of thiosulfate to reach the endpoint. What is the DO content of the sample?

DATA COLLECTION SHEET

DATA COLLECTION SHEET

DATA COLLECTION SHEET

DATA COLLECTION SHEET

19

DETERMINATION OF THE BIOCHEMICAL OXYGEN DEMAND OF SEWAGE INFLUENT

Purpose: To determine the biochemical oxygen demand in a domestic wastewater sample

To learn the Thomas slope method for determining the biochemical oxidation rate constant, k

BACKGROUND

The focus of this laboratory exercise will be to determine the amount of oxidizable organic matter (sewage) in a wastewater sample. As we discussed in the Chapter 18, the term *DO* refers to the chemical measurement of how much dissolved oxygen is present in a water sample, expressed in mg/L. The *biochemical oxygen demand* (BOD) is an estimate of how much total DO is required to oxidize the organic matter in a water sample. Thus, we will actually be measuring the change in DO in our experiments to estimate the BOD originally present in the water. But before we discuss the details of this experiment, it is important to gain an appreciation for the extent of the global sewage problem and environmental issues surrounding wastewater.

Our standard of living in the United States is a direct result of having adequate water and wastewater treatment, which are distinguishing features of developed

Environmental Laboratory Exercises for Instrumental Analysis and Environmental Chemistry
By Frank M. Dunnivant
ISBN 0-471-48856-9   Copyright © 2004 John Wiley & Sons, Inc.

countries. As early as 1700 B.C.E., people began to obtain the luxury of running water and then to deal with the disposal of associated wastes. Although there is evidence of plumbing and sewage systems at many age-old sites, including the *cloaca maxiumn*, or great sewer, of the ancient Roman empire, the common use of sewer and plumbing systems did not become widespread until modern times (Wastewater and Public Health, 2000). Along with providing drinking water and disposing of sewage comes the challenge of preventing the rapid spread of disease within populations that utilize a common water source and treatment facilities.

The microorganisms associated with waterborne diseases are (1) bacteria responsible for typhoid fever, cholera, and shigellosis; (2) viruses causing hepatitis and viral gastroenteritis; and (3) protozoans that are the agents of the waterborne diseases cryptosporidiosis and giardiasis. Due to the importance of these diseases in our history and in the current state of many undeveloped countries, a brief description of the organisms and symptoms will be given.

Early in this century, typhoid fever was the most commonly reported waterborne disease in the United States. Tartakow and Vorperian (1981) cite in their study *Food and Waterborne Diseases: Their Epidemiologic Characteristics* that in 1900 there were as many as 350,000 cases of typhoid fever causing 35,000 deaths. The bacterium *Salmonella typhi* causes this disease by invading the intestine, replicating in lymph nodes, and entering the bloodstream. The symptoms of typhoid fever can take from one to several weeks to appear and include fever, malaise, headache, the inability to eat, a transient rash, and diarrhea or constipation. The mortality rates are as high as 12 to 16% but drop to as low as 1 to 4% if treated with antibiotics (Craun, 1986). One of the largest waterborne typhoid fever outbreaks in the United States occurred in Riverside, California, in 1965, where a contaminated municipal water system resulted in 16,000 cases, with 20 people hospitalized and three deaths (Nauman, 1983). The largest waterborne cholera outbreak in the United States since the beginning of the twentieth century occurred in 1981, when cholera was diagnosed in four cases of severe diarrhea caused by sewage contamination of the private water system of an oil rig (Craun, 1986). In developed countries, cholera has been controlled by protection and purification of the water supply with chlorination and with the sanitary disposal and treatment of sewage. However, in undeveloped countries cholera still claims lives each year.

Shigellosis or bacillary dysentery is caused by four pathogenic species: *Shigella boydii, S. dysenteriae, S. flexneri*, and *S. sonnei*, but it is not commonly infectious by small amounts in fecal–oral transmission. Most of the outbreaks of shigellosis are caused by person-to-person infections (Craun, 1981). The first reported cases in the United States were not until between 1936 and 1945, with no associated deaths. The symptoms of this disease are infection of the bowel with diarrhea, and often with traces of blood or mucus, fever, cramps, and vomiting. No more than 6% of the cases of shigellosis reported in the United States have occurred because of waterborne outbreaks, and only 2% were caused by drinking water (Craun, 1981). The vast majority of the waterborne outbreaks were due to the contamination of untreated groundwater and the lack of adequate disinfection for groundwater. Treatment systems and natural water bodies are not regularly monitored unless contamination is suspected.

Viral hepatitis, mostly hepatitis A, is the second most commonly reported disease in waterborne outbreaks in the United States (Craun, 1986). Hepatitis was mentioned early in history in the writings of the Greek physicians of the second century as epidemic jaundice (Tartakow and Vorperian, 1981). Between the years 1629 and 1868, 53 European and 11 American cities reportedly experienced outbreaks of hepatitis (Tartakow and Vorperian, 1981). The symptoms of hepatitis A include nausea, vomiting, diarrhea, malaise, abdominal discomfort, weakness, and fever, followed by jaundice and brownish urine (Craun, 1986). The disease spreads through the fecal–oral route and may last as long as several months. Viruses can be removed effectively from groundwater and surface water by adsorption and filtration along with chlorination and ozonation.

Giardiasis has been recognized as a pathogen only since the early 1980s, when it was also identified as the most commonly detected intestinal parasite in the United States and the United Kingdom (Nauman, 1983). Giardiasis is an infection of the small intestine caused by *Giardia lamblia* with a range of symptoms from an asymptomatic cyst passage state to severe gastrointestinal involvement, resulting in diarrhea, abdominal cramps, fatigue, and weight loss. Giardiasis may be spread by feces from a carrier, sewage contaminating a water supply, or by hand-to-mouth transfer of cysts (Tartakow and Vorperian, 1981). The main concern with giardiasis is that disinfection with chlorine does not adequately treat most water for *Giardia* cysts. For complete protection, water treatment should include sedimentation, filtration, and chlorination (Nauman, 1983). Along with giardiasis, cryptosporidiosis has been one of the most recent waterborne pathogens associated with water treatment. Cryptosporidiosis is caused by the one-celled parasite *Cryptosporidium*. Profuse diarrhea, headaches, fever, cramping, and nausea characterize illness caused by this parasite. The disease is spread like giardiasis, by the fecal–oral route, person-to-person contact, animal-to-person contact, ingestion of contaminated water or food, or hand-to-mouth contact with the oocysts. The disinfection of surface waters from this disease has proven problematic because the oocysts are strongly resistant to chlorine. Filtration, in some cases, and ozonation have been proven effective, together with specific chlorination processes, in reducing the risk of water with the *Cryptosporidium* parasite.

While the diseases discussed above pose risks to human populations, the release of untreated sewage to waterways can also result in the immediate death of aquatic systems. Surface aquatic systems are aerobic, and the life-forms contained in these systems are dependent on the constant presence of dissolved oxygen (DO). Most streams are at or near saturation with respect to DO, and when readily oxidizable organic matter such as domestic sewage enters the stream, native microorganisms not only rapidly consume all of the DO present but consume oxygen faster than it can be replenished through reaeration. The DO in waterways

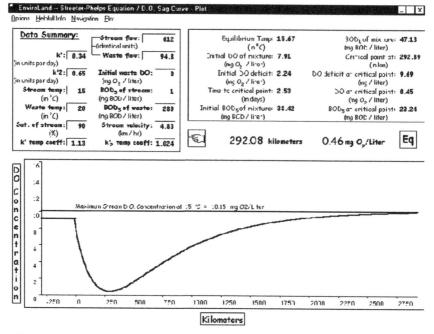

Figure 19-1. Streeter–Phelps plot of the DO in a stream receiving domestic sewage.

downstream from a sewage plant can be modeled with the *Streeter–Phelps equation*,

$$D = \frac{k' \cdot \text{BOD}_L}{k_2' - k'} \left(e^{-k'(x/v)} - e^{-k_2'(x/v)}\right) + D_0 e^{-k_2'(x/v)}$$

where
D = dissolved oxygen concentration (mg O_2/L)
k' = BOD rate constant for oxidation (day^{-1})
BOD_L = ultimate BOD (mg/L)
k_2' = reaeration constant to the base e (day^{-1})
x = distance from the point source (miles or kilometers)
v = average water velocity (miles/day or kilometers/day but units must be compatible with distances, x)
D_0 = initial oxygen deficit (mg/L) of the stream (saturated value minus the actual DO concentration)

This equation is discussed in detail in Chapter 28. A typical plot from this equation for a stream receiving raw sewage is shown in Figure 19-1. Note the shape of the curve. Initially, above the entry point of the sewage, the water is near saturation with respect to DO. As sewage enters a stream, the DO concentration plummets to near zero and often does drop to zero. As the organic matter is oxidized and the stream reaerates, the DO level slowly rises, achieving background natural concentration. Figure 19-2 shows the dramatic effect of

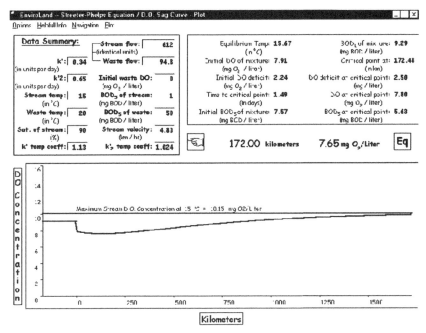

Figure 19-2. Streeter–Phelps plot of the DO in a stream receiving domestic sewage, but with a wastewater treatment plant online.

treating (or not treating) the sewage before discharging it into the stream. Above the sewage entry point, the water is near saturation with respect to DO, and it drops drastically after the sewage enters the stream without treatment.

The focus of modern sewage treatment is to remove turbidity, readily oxidizable organic matter, and pathogenic organisms. These three goals can easily be achieved at a minimal cost. Turbidity is removed in primary and secondary clarifiers and in sand bed filters. Organic matter is removed in biological contact units such as trickling filters and activated sludge lagoons. Most pathogens are naturally removed in the various treatment process, but removal is ensured with the use of sand bed filtration, chlorination, and ozonation. One of the major design criteria for a wastewater treatment plant, and in fact a daily monitoring parameter, is the biochemical oxygen demand of the incoming and outgoing waste. In this laboratory exercise, we measure the five-day BOD (BOD_5), the ultimate BOD (BOD_L), and the microbiological oxidation rate (k).

THEORY

In general, the utilization of oxygen by microorganisms is considered to be a pseudo-first-order process which for a closed system (no reaeration) is commonly described by

$$L = L_0 e^{-kt} \tag{19-1}$$

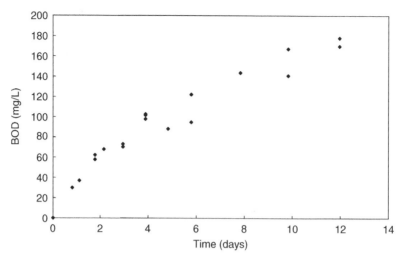

Figure 19-3. BOD data illustrating the exponential trend in oxygen depletion.

where L is the concentration of oxygen at time t, L_0 the original concentration of oxygen in a sample, k the rate constant (generally around 0.17 day^{-1} for sewage waste), and t is time. Equation (19-1) is used to draw the line representing the removal of oxygen illustrated in Figure 19-3.

A similar expression can be used to describe the oxidation of BOD in the sample since it is the inverse of the oxygen consumption,

$$L = L_0 - L_0 e^{-kt} \tag{19-2}$$

where L is the concentration of biodegradable organic matter at time t, L_0 the original or ultimate concentration of biodegradable organic matter, k the rate constant (generally around 0.17 day^{-1} for sewage waste), and t is time.

Traditionally, we are concerned with the amount of oxygen required to oxidize the BOD over a five-day period. This time period was established years ago in England and results from the fact that it requires five days for the water in most English streams to reach the ocean. The BOD continues to exert an oxygen demand on the stream after this time, and the ultimate BOD determined over a 20-day period is becoming commonly used in the United States. The ultimate BOD, L_0, can be determined using the *Thomas slope method* (Snoeyink and Jenkins, 1980), which linearizes the data in the form

$$\left(\frac{t}{y}\right)^{1/3} = (L_0 k)^{-1/3} + \frac{k^{2/3}}{6L_0^{1/3}} t \tag{19-3}$$

where t is the time, y the BOD in mg/L at time t [L in equation (19-2)], L_0 the original or ultimate concentration of biodegradable organic matter, and k the rate constant.

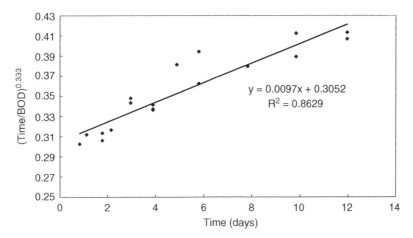

Figure 19-4. Thomas slope analysis of BOD data.

An alternative method of determining L_0 is to measure the BOD over a five-day period, fit the data to equation (19-2) using a k of 0.17, and solve for L_0. However, experience has shown that this method does not work well, due to the non-first-order nature of the microbial degradation process. Data from Figure 19-3 have been transformed according to equation (19-3) and are shown in Figure 19-4. You will undoubtedly note the scatter in the data plot. This is common in BOD experiments, where we are dependent on microbial reactions and growth rates that are not highly reproducible.

Note that equation (19-3) is the equation of a line, where

$$(L_0 k)^{-1/3} = y \text{ intercept, } b$$

$$\frac{k^{2/3}}{6L_0^{1/3}} = \text{slope of the line,} m$$

By substitution, it can be shown that $k = 6m/b$ and $L_0 = 1/kb^3$. Thus, by plotting an experimental data set of lab measurements (BOD as a function of time; Figure 19-3) according to equation (19-3), the rate constant and ultimate BOD can be estimated. For the data in Figure 19-4, this yields a rate constant, k, of 0.191 day^{-1} and a BOD$_L$ of 184 mg/L.

ACKNOWLEDGMENTS

The epidemiologic discussion in the introduction was contributed by Celeste Bolin (Whitman College, Class of 2000). Data presented in this chapter were collected by the Whitman environmental chemistry class of 2000.

REFERENCES

American Water Works Association, *Standard Methods for the Examination of Water and Wastewater*, 20th ed., AWWA, Denver, CO, 1998.

Craun, G., *Waterborne Diseases in the United States*, CRC Press, Boca Raton, FL, 1986.

Nauman, H., thesis submitted to Oregon State University, 1983.

Sawyer, C. N. and P. L. McCarty, *Chemistry for Environmental Engineering*, McGraw-Hill, New York, 1978.

Snoeyink, V. L. and D. Jenkins, *Water Chemistry*, Wiley, New York, 1980.

Taktakow, I. J. and J. H. Vorperian, *Food and Waterborne Diseases: Their Epidemiologic Characteristics*, AVI Publishing, Westport, CT, 1981.

Wastewater and Public Health, `http://danpatch.ecn.purdue.edu/~epados/septics/disease.htm`, accessed Feb. 24, 2000.

IN THE LABORATORY

A BOD determination is made by taking a sample and incubating it over a five- or 20-day period and monitoring the dissolved oxygen concentration at intervals of 12 or more hours. For high concentrations of BOD, the sample must be diluted to avoid depleting all of the original oxygen present in the water sample. There are several requirements for the dilution water. For example, pure distilled water should not be used since microorganisms require certain salts for proper metabolism. Thus, potassium, sodium, calcium, magnesium, iron, and ammonium salts are added to the dilution water. Also, the water's pH should be buffered between 6.5 and 8.5 with phosphate buffers. Some water samples require a "seed" of viable microorganisms to complete the degradation process. A general rule of thumb has been developed to provide sufficient accuracy in determining BOD values. This states that at least 2 mg/L of oxygen must be used over the course of the experiment (five or 20 days), but at least 0.5 mg/L must remain in the final sample. The oxygen concentration can be measured by one of two methods described in Chapter 18.

Safety Precautions

- As in all laboratory exercises, safety glasses must be worn at all times.
- Avoid skin and eye contact with caustic and acidic solutions. If contact occurs, rinse your hands and/or flush your eyes for several minutes. Seek immediate medical advice for eye contact.
- Use concentrated acids in the fume hood and avoid breathing their vapors.

Chemicals and Solutions

- Phosphate buffer solution. Dissolve 8.5 g of KH_2PO_4, 21.75 g of K_2HPO_4, 33.4 g of $Na_2HPO_4 \cdot 7H_2O$, and 1.7 g of NH_4Cl in approximately 500 mL deionized water and dilute to 1.0 L. The pH of this solution should be 7.2.
- Magnesium sulfate solution. Dissolve 22.5 g of $MgSO_4 \cdot 7H_2O$ in deionized water and dilute to 1.0 L.
- Calcium chloride solution. Dissolve 27.5 g of $CaCl_2$ in deionized water and dilute to 1.0 L.
- Ferric chloride solution. Dissolve 0.25 g of $FeCl_3 \cdot 6H_2O$ in deionized water and dilute to 1.0 L.
- Acid and alkali solutions for adjusting wastewater samples with extreme pH values. 1 M HCl and 1 M NaOH.
- Sodium sulfite solution for removal of residual chlorine. Dissolve 1.575 g of Na_2SO_4 in 1.0 L of deionized water. This solution is not stable and should be prepared daily.

Equipment and Glassware

- BOD incubator or a 20°C water bath that can be kept in the dark
- 250- or 300-mL BOD bottles
- 1-mL wide-bore pipets for pipetting the blended sewage mixture
- Several 1.00-mL pipets
- All of the glassware needed in the Winkler titration experiment

PROCEDURES

Preparation of Dilution Water

You will be determining the BOD of a domestic wastewater (sewage) sample, and you will need to dilute the wastewater in order to monitor the consumption of DO over a five- or 20-day period. Since you will be diluting the wastewater, it is necessary to add inorganic nutrients to the dilution water. The nutrients include the phosphate buffer, calcium chloride, and ferric chloride solutions described earlier in the "Chemicals and Solutions" section. First, estimate how much dilution water you will need. In doing this you must determine the dilution factor of your wastewater and the number of replicates you will be taking. A rule of thumb for estimating the dilution factor of your wastewater can be determined from Table 19-1. The best way to determine the appropriate dilution factor is to consult the operator of the sewage treatment plant where you obtain your wastewater sample. After you have estimated the volume of dilution water needed, add 1 mL of each nutrient solution (phosphate buffer, $MgSO_4$, $CaCl_2$, and $FeCl_3$) per liter of dilution water needed. It is best if you store the dilution water at $20°C$ overnight to allow O_2 equilibration with the atmosphere.

Seeding

A *seed* is needed when your sample does not have sufficient microbial community to support exponential microbial growth immediately. A seed usually consists of a small amount of sewage added to your samples. If you are using domestic wastewater, you will probably not have to seed your water, since viable microbial

TABLE 19-1. BOD Measurable from Various Dilutions of Sample

Using Percent Mixtures		By Direct Pipetting into 300-mL Bottles	
% Mixture	Range of BOD (mg/L)	mL Sample	Range of BOD (mg/L)
0.01	20,000–70,000	0.02	30,000–105,000
0.02	10,000–35,000	0.05	12,000–42,000
0.05	4,000–14,000	0.10	6,000–21,000
0.1	2,000–7,000	0.20	3,000–10,500
0.2	1,000–3,500	0.50	1,200–4,200
0.5	400–1,400	1.0	600–2,100
1.0	200–700	2.0	300–1,050
2.0	100–350	5.0	120–420
5.0	40–140	10.0	60–210
10.0	20–70	20.0	30–105
20.0	10–35	50.0	12–42
50.0	4–14	100	6–21
100	0–7	300	0–7

Source: AWWA (1998).

communities are already present. For the purposes of this experiment we will assume that you do not need to seed your samples, but keep in mind that river, lake, and groundwater samples often need to have a seed added for BOD determination. When you do use a seed, you must also run a blank for your BOD determination, since the seed will consume a small amount of the DO.

pH Adjustment

Some domestic wastewater samples have industrial inputs to the sewer system and as a result may have extreme pH values (very high or low). In these cases it will be necessary to adjust the pH of your original wastewater sample prior to making dilutions according to Table 19-1. Use 1 M HCl or 1 M NaOH for these adjustments.

Chlorine Removal

Some samples may contain residual chlorine compounds that will inhibit the growth of microorganisms and will interfere with the BOD determination. If your sample contains residual chlorine compounds, these can be removed with sodium sulfite. Domestic sewage samples rarely have residual chlorine compounds and we will not use sulfite in this procedure, but be aware that this is not always the case.

Setup and Titration of BOD Samples

1. Determine the appropriate dilution of your wastewater based on Table 19-1 and/or data from the wastewater treatment plant operator. It is best to have three dilutions, one 20 to 30% less dilute than suggested, one as suggested, and one 20 to 30% more concentrated than suggested. This approach should allow determination of the BOD_5 and BOD_L.

2. Before you make your dilutions, homogenize your wastewater sample by blending it in a food processor at high speed for 5 minutes. Also adjust the temperature to 20°C.

3. Add the desired volume of BOD to each BOD bottle and fill with equilibrated, nutrient-added, 20°C dilution water. (Alternatively, you may mix a larger volume of wastewater-dilution water and fill your BOD bottles.)

4. Make sure that the bottles are filled to the top with dilution water. Insert the tapered cap in a manner to exclude any air bubbles from the BOD bottle. Titrate two samples initially to determine the DO at $t = 0$.

5. Incubate the dilutions at 20°C for 20 days, taking bottles from each dilution at each time interval (based on Figure 19-3) and titrating them using the

Winkler method to obtain a plot of BOD versus time. The necessary sampling times are dependent on the microbial oxidation rate, k, but you can use the data in Figure 19-3 as an initial estimate of sampling times.

6. Refer to the Winkler method (Chapter 18) for the fixing and DO titration procedures.

7. Analyze your data and determine the BOD_5, BOD_L, and microbial rate constant.

Waste Disposal

After neutralization, all solutions can be disposed of down the drain with water.

ASSIGNMENT

For this lab exercise you will write a formal lab report. This will consist of:

- A brief introduction about BOD or sewage waste with a few references from the Internet or library
- A procedure for making the dilution water and the solutions
- A procedure for the titrations [neither this nor the procedure above need details on how to make the stock solutions other than to reference the lab instructions or AWWA (1998)]
- A summary table of all of the BOD titration data, with the DO and BOD in mg/L
- A Thomas slope plot and calculations
- A summary/conclusion section

Refer to each plot by a figure number. Your instructor may ask you to use the Streeter–Phelps simulator included on the CD-ROM (in Fate) to model the effect of the treatment plant on water quality. Simulate two scenarios, one where the treatment plant fails and all of the incoming sewage is input directly to the receiving stream, and another where the treatment plant removes 99% of the incoming BOD. How does the treatment plant improve water quality in the stream?

ADVANCED STUDY ASSIGNMENT

Discuss the following prelab questions for the BOD laboratory.

1. What is the difference between DO and BOD?
2. When do you need to seed BOD samples?
3. If you have a sewage sample estimated to contain 5000 mg/L BOD, at what dilution would you recommend running the BOD test?

DATA COLLECTION SHEET

DATA COLLECTION SHEET

DATA COLLECTION SHEET

DATA COLLECTION SHEET

DATA COLLECTION SHEET

20

DETERMINATION OF INORGANIC AND ORGANIC SOLIDS IN WATER SAMPLES: MASS BALANCE EXERCISE

Purpose: To develop your weighing and laboratory skills

To learn the concept of a mass balance

BACKGROUND

Mass balances (an accounting of all mass of a pollutant in a defined system) are important concepts in environmental chemistry and geochemistry. Mass balances can be conducted on any element or compound but are usually illustrated in the classroom using global mass balances of the water, nitrogen, sulfur, carbon, and phosphorus cycles. Examples of these can be found in Berner and Berner (1996). In this laboratory exercise, we will collect data and conduct a mass balance on inorganic and organic solids in a water sample. Due to the complexity of this experiment and time constraints, the class will be divided into three groups, with each group conducting a different experiment. Thus, everyone will have to keep careful records and share data with the rest of the class. But first we will answer two questions concerning suspended and dissolved solids in typical water samples:

1. Why are we concerned with total suspended solids (TSS)?

    ○ High concentrations of suspended solids may settle out onto a streambed or lake bottom and cover aquatic organisms, eggs, or macroinvertebrate

Environmental Laboratory Exercises for Instrumental Analysis and Environmental Chemistry
By Frank M. Dunnivant
ISBN 0-471-48856-9   Copyright © 2004 John Wiley & Sons, Inc.

larva. This coating can prevent sufficient oxygen transfer and result in the death of buried organisms.

o High concentrations of suspended solids decrease the effectiveness of drinking water disinfection agents by allowing microorganisms to "hide" from disinfectants within solid aggregates. This is one of the reasons that the TSS, or turbidity, is removed in drinking water treatment facilities.

o Many organic and inorganic pollutants sorb to soils so that the pollutant concentrations on the solids are high. Thus, sorbed pollutants (and solids) can be transported elsewhere in river and lake systems, resulting in the exposure of organisms to pollutants away from the point source.

2. Why are we concerned with total dissolved solids (TDS)?

o The total dissolved solids (TDS) of potable waters range from 20 to 1000 mg/L. In general, waters with a TDS value below 500 mg/L are most desirable for domestic use.

o This is because waters with TDS > 500 mg/L may cause diarrhea or constipation in some people.

o Water with a high TDS value is frequently hard (i.e., has a high Ca^{2+} and Mg^{2+} concentration) and requires softening (the removal of hardness cations) by precipitation.

o Waters with high TDS value may result in clogged pipes and industrial equipment through the formation of scale (Ca and Mg precipitates).

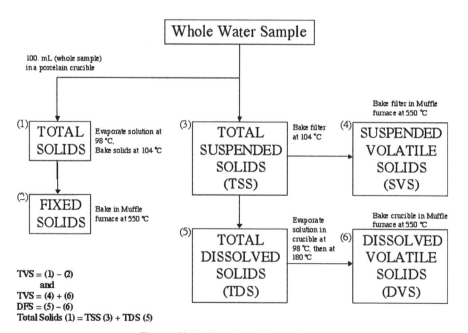

Figure 20-1. Overview of procedures.

The TDS value of a water sample can be used to determine the most appropriate method of water softening, since precipitation reduces TDS, while some ion exchange processes may increase TDS.

THEORY

This laboratory exercise will test your analytical, technique, and logic abilities. First you must read and organize the attached procedures for each experiment. These are compiled from the *Standard Methods for the Examination of Water and Wastewater* (American Public Health Association, 1992, procedures 2540A–E) and were combined to create Figure 20-1. A sample of known composition will be provided, containing both inorganic and organic solids.

REFERENCES

American Public Health Association, *Standard Methods for the Examination of Water and Waste Water*, APHA, Washington, DC, 1992.

Berner, E. K. and R. A. Berner, *Global Environment: Water, Air, and Geochemical Cycles*, Prentice Hall, Upper Saddle River, NJ, 1996.

Sawyer, C. N., P. L. McCarty, and G. F. Parkin, *Chemistry for Environmental Engineering*, McGraw-Hill, New York, 1994.

Snoeyink, V. L. and D. Jenkins, *Water Chemistry*, Wiley, New York, 1980.

IN THE LABORATORY

Equipment and Glassware

For the entire class, three groups:

- 1.00 L of unknown sample
- 8 to 10 100-mL or larger porcelain crucibles (high-silica crucibles may also be used)
- 10 to 15 glass-fiber filters (1 μm nominal size, without organic binders) and filtration setups
- Five filtration flasks (500-mL capacity)
- Three graduated cylinders (100-mL capacity)
- 98°C oven
- 104°C oven
- 180°C oven
- Muffle furnace (550°C)
- Six desiccators

PROCEDURE

Note: Each of the procedures described here will require you to come in at unusual times during the next week. You must observe the safety procedures set up by your school pertaining to working in the laboratory alone or in pairs.

Group I: Total (TS), Fixed (FS), and Volatile Solids (VS) (Boxes 1 and 2 in Figure 20-1)

Overview. In this procedure you will be taking 100 mL of the entire water sample and adding it to a crucible. In case you do not have a crucible that will hold 100 mL, you can add 50 mL, evaporate the sample to dryness, and add another 50-mL aliquot. You will first evaporate the sample to dryness in a 98°C oven to avoid boiling or splattering of the sample that could result in loss of the sample. After all of the water has evaporated, you will bake the crucible at 104°C to remove all or most of the water (note that occluded water may remain). The difference between the initial weight of the crucible (heated to 104°C, cooled, and empty) and the weight after adding sample (heated to 104°C), divided by the total volume of sample you added to the crucible, will yield the total solids in your sample. This is the value for box 1 in Figure 20-1. Next you will determine the fixed solids of your sample by taking the 104°C dried crucible and heating it to 550°C. This heating will oxidize all of the organic carbon in your sample. You will obtain this number by taking the difference between the 104°C and the 550°C weights, divided by the total volume of your sample (100 mL).

Step-by-Step Instructions

Preparing the Crucibles

1. Wash and clean at least three crucibles. You will only need three crucibles but some may crack in the preparation steps, so it is wise to have more prepared than needed. Some residue may remain from previous experiments.
2. Label each crucible if they are not already labeled. A good way to label crucibles is to paint a number on the bottom using iron oxide paste. When you bake the crucible, the iron will bind permanently to the crucible. After air-drying them thoroughly, dry the crucibles in the muffle furnace at 550°C for 30 minutes.
3. Obtain a constant weight for the crucibles by drying at 550°C, allowing the crucible to cool almost completely on a benchtop and then to cool completely in a desiccator, weighing to a constant weight, and repeating the heating and cooling process until you have two weights within 0.6 mg or less of each other.

Obtaining Your Total Solids Measurement

4. Add 100 mL of your sample to your crucible and evaporate it in the 98°C oven overnight.

5. Place your crucible in the 104°C oven the next day and bake it for 1 hour.

6. Place the crucible in a desiccator until cool and obtain a constant weight.

7. Repeat steps 5 and 6 until you obtain a constant weight (within 0.5 mg of each other). This will be the total solids measurement (box 1 in Figure 20-1):

TS(mg/L)

$$= \frac{(\text{average final weight in g} - \text{average initial crucible weight empty in g})(1000\,\text{mg/g})}{\text{sample volume in L}}$$

Obtaining Your Fixed and Volatile Solids Measurements

8. Place the crucible from step 7 in the muffle furnace at 550°C for 30 minutes.

9. Obtain a constant weight for the crucibles by allowing the crucibles to cool almost completely on a benchtop and then to cool completely in a desiccator, weighing to a constant weight, and repeating the heating and cooling process until you have two weights within 0.6 mg or less of each other. This measurement will yield your fixed solids and volatile solids measurements (box 2 in Figure 20-1).

FS(mg/L)

$$= \frac{(\text{average final weight from step 9 in g} - \text{average initial crucible weight empty in g})(1000\,\text{mg/g})}{\text{sample volume in L}}$$

VS(mg/L)

$$= \frac{(\text{average initial crucible weight from step 7 in g} - \text{average final weight from step 9 in g})(1000\,\text{mg/g})}{\text{sample volume in L}}$$

Group II: Total Suspended Solids (TSS) and Suspended Volatile Solids (SVS) (Boxes 3 and 4 in Figure 20-1)

Overview. In this procedure you will be taking 100 mL of your sample and performing the most commonly used solids measurement, the total suspended solids. This requires you to filter a known volume of sample through a preheated and pretared glass-fiber filter. The difference in weights (final–initial) divided by the volume of sample will yield the TSS. The TSS measurement accounts for all solids that do not pass through the filter (typically, 0.45 to 1 μm in size), weighed after drying at 104°C. When the filter is further dried to 550°C, you will oxidize any organic matter present in the solids and can obtain a suspended volatile solids measurement. Thus, you will be completing boxes 3 and 4 in Figure 20-1.

Step-by-Step Instructions

Preparing the Filters

1. Rinse three filters with 20 to 30 mL of deionized water to remove any solids that may remain from the manufacturing process. Place each filter in a separate, labeled aluminum weight pan, dry them in a 550°C muffle furnace for 30 minutes, place them (filter and pan) in a desiccator, and obtain a constant weight by repeating the oven and desiccation steps.

Obtaining the TSS Measurement

2. Filter 100 mL of sample through each filter.
3. Place each filter paper in the aluminum weight pan in the 104°C oven for 1 hour. Cool the filter and pan in a desiccator and obtain a constant weight by repeating the drying and desiccation steps. This procedure will yield the TSS measurement (box 3 in Figure 20-1):

$$TSS(mg/L)$$
$$= \frac{(\text{average weight from step 3 in g} - \text{average inital weight from step 1 in g})(1000\,mg/L)}{\text{sample volume in L}}$$

Obtaining the SVS Measurement

4. Place the filters (in the pan) from step 3 in a muffle furnace at 550°C for 30 minutes. (Depending on the type of muffle furnace you have, you may need to cover the samples with another weigh pan to avoid contamination of your sample by ceramic dust.) Remove the filters, place them in a desiccator, and obtain a constant weight by repeating the muffling and desiccation steps. This will yield the SVS measurement (box 4 in Figure 20-1):

$$SVS(mg/L)$$
$$= \frac{(\text{average weight from step 3 in g} - \text{average weight from step 4 in g})(1000\,mg/L)}{\text{sample volume in L}}$$

Group III: Total Dissolved Solids (TDS) and Dissolved Volatile Solids (DVS) (Boxes 5 and 6 in Figure 20-1)

Overview. In this procedure you will perform the second-most-common solids measurement, the total dissolved solids. This is determined by first performing a TSS measurement, but you do not have to weigh the filter as Group II had to do.

You are only concerned with removing the filterable solids from your sample and collecting the filtrate. In case you do not have a crucible that will hold 100 mL, you can add 50 mL of the filtrate, dry the sample to dryness, and add another 50-mL aliquot. You will first evaporate the sample filtrate to dryness in a 98°C oven, to avoid boiling or splattering, which could result in a loss of sample. Next, you will take the dried filtrate and crucible and dry it at 104°C to obtain the mass of dissolved solids in your sample. Finally, you will bake the sample at 550°C to oxidize the organic matter and obtain a dissolved volatile solids measurement.

Step-by-Step Instructions

Preparing the Crucibles

1. Wash and clean at least three crucibles. You only need three crucibles, but some of these may crack in the preparation steps, so it is wise to have more prepared than are needed. Some residue may remain from previous experiments.
2. Label each crucible if they are not already labeled. A good way to label crucibles is to paint a number on the bottom using an iron oxide paste. When you bake the crucible, the iron will bind permanently to the crucible. Allow the crucibles to dry completely prior to placing them in the muffle furnace. After air drying, dry the crucibles in the muffle furnace at 550°C for 30 minutes (the 30 minutes refers to the total time at 550°C).
3. Obtain a constant weight of the crucibles by drying at 550°C, allow the crucible to cool almost completely on a benchtop and then to cool completely in a desiccator, weigh to a constant weight, and repeat the heating and desiccation processes until you have two weights within 0.6 mg or less of each other.

Obtaining the Total Dissolved Solids Measurement

4. Add 100 mL of your *filtered* sample to your crucible and evaporate it in the 98°C oven overnight.
5. Place your crucible in the 104°C oven the next day and bake it for 1 hour.
6. Place the crucible in a desiccator until cool and obtain a constant weight.
7. Repeat steps 5 and 6 until you obtain a constant weight (within 0.5 mg of each other). This will be the total dissolved solids measurement (box 5 in Figure 20-1):

$$TDS(mg/L)$$
$$= \frac{(\text{average crucible weight from step 7 in g} - \text{average crucible weight empty in g})(1000\,mg/g)}{\text{sample volume in L}}$$

Obtaining the Volatile Dissolved Solids Measurements

8. Place the crucible from step 7 in the muffle furnace at 550°C for 30 minutes.

9. Obtain a constant weight of the crucibles by allowing the crucible to cool almost completely on a benchtop and then to cool completely in a desiccator, weighing to a constant weight, and repeating the heating and desiccation processes until you have two weights within 0.6 mg or less of each other. This measurement will yield your total volatile solids measurements (box 6 in Figure 20-1):

$$DVS(mg/L)$$
$$= \frac{(\text{average final crucible weight from step 7 in g} - \text{average crucible weight from step 9 in g})(1000 \, mg/g)}{\text{sample volume in L}}$$

Additional Procedure

If you have a conductivity meter, you will also be measuring the conductivity of the entire water sample. The conductivity of a sample is a measure of the dissolved salt concentration (TDS), but the conductivity depends on type of ions (monodivalent, divalent, etc.), their concentration, and the temperature. In general,

$$TDS = \text{conductivity (mmhos/cm)} \times K$$

where K is a constant ranging from 0.55 to 0.90, depending on the ions in solution and the temperature. Calibrate the conductivity meter using the 0.00100 M KCl solution. This solution should have a conductivity of 146.9 mS/cm, and the 0.00500 KCl solution should have a conductivity of 717.5 mS/cm at 25°C. Make sure that the meter's automatic temperature compensation function is turned on.

1. Measure and record the conductivity of the sample. If the conductivity of the sample is nearer to 700 mS/cm than to 150 mS/cm, recalibrate the meter with the higher concentration standard.

2. Change the meter from conductivity to TDS mode and measure the TDS of your sample. Note that this meter uses a K value of 0.5 to estimate TDS from conductivity. This assumes that the only ions present in solution are Na and Cl. The TDS value reported by the meter has units of mg/L.

3. Compare your conductivity value to your measured TDS value.

If you have a turbidity meter, you will also be measuring the turbidity of your original sample. Turbidity is a measure of the light scattered by suspended particles, especially the clay particles in the sample. Turbidity is measured by a photoelectric detector aligned at a 90° angle from the light source. Turbidity, measured in nephelometer turbidity units (NTUs), is a function of particle size, shape, and concentration. Turbidity is only a quick field approximation of total suspended solids. Consult the user's manual and measure the turbidity of your sample. Compare this to your TSS measurement.

Hints for Success

- Always, *always* mix your sample completely before removing any solution or suspension. The clay particles will settle and bias your results if you do not mix the sample completely every time you remove an aliquot.

- Normally in this laboratory manual you will not be given data collection sheets, but due to the complicated nature of this experiment, a data sheet is supplied on your CD. Each group should complete the data sheet for the experiments that you conduct and share the results with your instructor and the other groups.

- Perform all measurements in triplicate.

- Carefully clean all containers and prewash all filters with deionized water prior to use. As the procedure section notes, heat all of these to the maximum temperature that you will use before obtaining weights. Also as noted in the procedure, you must obtain a constant weight (generally within 0.5 mg of each other) before you end each experiment. (Fingerprints and dust weigh enough to affect your results significantly.)

- Your balances have been calibrated, but for best results you should use the same balance for every measurement. Even if the calibration on a balance is slightly off, the change in weight will probably be accurate.

ASSIGNMENT

Complete the data sheet included with this laboratory procedure. In addition to the TS, FS, VS, SVS, and DVS calculations, you should answer the questions located on the bottom of the data collection sheet on the enclosed CD. Based on your results, also summarize the mass of particulate matter, inorganic salts, and organic matter in your 1.00-L sample.

DATA COLLECTION SHEET

DATA COLLECTION SHEET

DATA COLLECTION SHEET

DATA COLLECTION SHEET

DATA COLLECTION SHEET

21

DETERMINATION OF ALKALINITY OF NATURAL WATERS

Purpose: To determine the alkalinity of a natural water sample by titration

BACKGROUND

Alkalinity is a chemical measurement of a water's ability to neutralize acids. Alkalinity is also a measure of a water's buffering capacity or its ability to resist changes in pH upon the addition of acids or bases. The alkalinity of natural waters is due primarily to the presence of weak acid salts, although strong bases may also contribute in industrial waters (i.e., OH^-). Bicarbonates represent the major form of alkalinity in natural waters and are derived from the partitioning of CO_2 from the atmosphere and the weathering of carbonate minerals in rocks and soil. Other salts of weak acids, such as borate, silicates, ammonia, phosphates, and organic bases from natural organic matter, may be present in small amounts. Alkalinity, by convention, is reported as mg/L $CaCO_3$, since most alkalinity is derived from the weathering of carbonate minerals rather than from CO_2 partitioning with the atmosphere. Alkalinity for natural water (in molar units) is typically defined as the sum of the carbonate, bicarbonate, hydroxide, and hydronium concentrations such that

$$[\text{alkalinity}] = 2[CO_3^{2-}] + [HCO_3^-] + [OH^-] - [H_3O^+] \qquad (21\text{-}1)$$

Alkalinity values can range from zero from acid rain–affected areas, to less than 20 mg/L for waters in contact with non-carbonate-bearing soils, to 2000 to

Environmental Laboratory Exercises for Instrumental Analysis and Environmental Chemistry
By Frank M. Dunnivant
ISBN 0-471-48856-9   Copyright © 2004 John Wiley & Sons, Inc.

4000 mg/L for waters from the anaerobic digestors of domestic wastewater treatment plants (Pohland and Bloodgood, 1963).

Neither alkalinity nor acidity, the converse of alkalinity, has known adverse health effects, although highly acidic or alkaline waters are frequently considered unpalatable. However, alkalinity can be affected by or affect other parameters. Below are some of the most important effects of alkalinity.

1. The alkalinity of a body of water determines how sensitive that water body is to acidic inputs such as acid rain. A water with high alkalinity better resists changes in pH upon the addition of acid (from acid rain or from an industrial input). We discuss this further when we discuss the relevant equilibrium reactions.

2. Turbidity is frequently removed from drinking water by the addition of alum, $Al_2(SO_4)_3$, to the incoming water followed by coagulation, flocculation, and settling in a clarifier. This process releases H^+ into the water through the reaction

$$Al^{3+} + 3H_2O \rightarrow Al(OH)_3 + 3H^+ \tag{21-2}$$

For effective and complete coagulation to occur, alkalinity must be present in excess of that reacted with the H^+ releases. Usually, additional alkalinity, in the form of $Ca(HCO_3)_2$, $Ca(OH)_2$, or Na_2CO_3 (soda ash), is added to ensure optimum treatment conditions.

3. Hard waters are frequently softened by precipitation methods using CaO (lime), Na_2CO_3 (soda ash), or NaOH. The alkalinity of the water must be known in order to calculate the lime, soda ash, or sodium hydroxide requirements for precipitation.

4. Maintaining alkalinity is important to corrosion control in piping systems. Corrosion is of little concern in modern domestic systems, but many main water distribution lines and industrial pipes are made of iron. Low-pH waters contain little to no alkalinity and lead to corrosion in metal pipe systems, which are costly to replacement.

5. Bicarbonate (HCO_3^-) and carbonate (CO_3^{2-}) can complex other elements and compounds, altering their toxicity, transport, and fate in the environment. In general, the most toxic form of a metal is its uncomplexed hydrated metal ion. Complexation of this free ion by carbonate species can reduce toxicity.

THEORY

As mentioned previously, alkalinity in natural water is due primarily to carbonate species. The following set of chemical equilibria is established:

$$CO_2 + H_2O \Leftrightarrow H_2CO_3^* \tag{21-3}$$

$$H_2CO_3 \Leftrightarrow HCO_3^- + H^+ \tag{21-4}$$

$$HCO_3^- \Leftrightarrow CO_3^{2-} + H^+ \tag{21-5}$$

where $H_2CO_3^*$ represents the total concentration of dissolved CO_2 and H_2CO_3. Reaction (21-3) represents the equilibrium of CO_2 in the atmosphere with dissolved CO_2 in the water. The equilibrium constant, using Henry's law, for this reaction is

$$K_{CO_2} = \frac{[H_2CO_3]}{P_{CO_2}} = 4.48 \times 10^{-5} \, M/mmHg \qquad (21\text{-}6)$$

The equilibrium expressions for reactions (21-4) and (21-5) are

$$K_1 = \frac{[H^+][HCO_3^-]}{[H_2CO_3]} = 10^{-6.37} \qquad (21\text{-}7)$$

$$K_2 = \frac{[H^+][CO_3^{2-}]}{[HCO_3^-]} = 10^{-10.32} \qquad (21\text{-}8)$$

As you can see from equations (21-6) to (21-8), the important species contributing to alkalinity are CO_3^{2-}, HCO_3^-, and H_2CO_3, and each of these reactions is tied strongly to pH. To illustrate the importance of these relations, we will calculate the pH of natural rainwater falling through Earth's atmosphere that currently contains 380 ppm CO_2.

First, we convert the concentration of CO_2 in the air to mol/L (step 1), and then calculate its partial pressure for use in equation (21-6) (step 2). This enables us to calculate the molarity of carbon dioxide in water [the $[H_2CO_3]$ term in equation (21-6)] (step 3), and then the molarity of H_2CO_3 in the water (step 4). Finally, we calculate the pH of the water, based on the equilibrium established between the different species of dissolved carbonate (step 4).

Step 1:

$$\text{density of air} = 0.001185 \, g/mL(1000 \, mL/L) = 1.185 \, g/L$$

$$CO_2(\text{air}) = 380 \, mg \, CO_2/kg \, \text{air}$$

$$= 380 \, mg \, CO_2/kg \, \text{air}(1 \, kg \, \text{air}/1000 \, g \, \text{air})(1.185 \, g/L)$$

$$= 0.450 \, mg \, CO_2/L$$

$$0.450 \, mg/L(1 \, g/1000 \, mg)(1 \, mol \, CO_2/44 \, g \, CO_2)$$

$$= 1.02 \times 10^{-5} \, M \, CO_2 \text{ in air}$$

Step 2:   Using $PV = nRT$ (note that $n/V = M$) gives us

$$P_{CO_2} = MRT = (1.02 \times 10^{-5} \, mol/L)(0.08206 \, L \cdot M/mol \cdot K)(298.14 \, K)$$

$$= 2.50 \times 10^{-4} \, atm$$

Step 3:   Using $K_{CO_2} = [CO_2]_{H_2O}/P_{CO_2} = 4.48 \times 10^{-5}\,M/\text{mmHg}$ yields

$$P_{CO_2}(\text{mmHg}) = 2.50 \times 10^{-4}\,\text{atm}\,(760\,\text{mmHg/atm}) = 0.19\,\text{mmHg}$$
$$K_{CO_2} = 4.48 \times 10^{-5}\,M/\text{mmHg} = M_{CO_2}/P_{CO_2}$$
$$M_{CO_2}\ \text{in water} = 4.48 \times 10^{-5}\,M/\text{mmHg}\,(0.19\,\text{mmHg})$$
$$= 8.52 \times 10^{-6}\,M\,CO_2$$

Step 4:   From step 3, $CO_2(\text{aq}) = 8.52 \times 10^{-6}\,M$;

$$CO_2(g) + H_2O \Leftrightarrow H_2CO_3 \qquad K = 1.88$$
$$K = \frac{[H_2CO_3]}{CO_2(\text{aq})}$$
$$[H_2CO_3] = 1.88(8.52 \times 10^{-6}\,M)$$
$$= 1.6 \times 10^{-5}\,M\,H_2CO_3$$

Step 5: Now, solving for pH using the equilibrium expression for H_2CO_3, we obtain

$$H_2CO_3 + H_2O \Leftrightarrow H_3O^+ + HCO_3^- \qquad K_a = 4.2 \times 10^{-7}$$
$$K_a = 4.2 \times 10^{-7} = \frac{[H_3O^+][HCO_3^-]}{[H_2CO_3]}$$
$$4.2 \times 10^{-7} = \frac{x^2}{1.6 \times 10^{-5} - x}$$

and using the quadratic equation to solve for x yields

$$x = [H_3O^+] = 2.59 \times 10^{-6}$$
$$\text{pH} = 5.59 \qquad \text{pH of natural rainwater}$$

We can also solve for the remaining chemical species using equilibrium equations.

$$[HCO_3^-] = x \text{ also, so}[HCO_3^-] = 2.59 \times 10^{-6}\,M\,HCO_3^-$$
$$[H_2CO_3] = 1.6 \times 10^{-5}(\text{total carbonic concentration})$$
$$- 2.59 \times 10^{-6} = 1.3 \times 10^{-5}\,M$$
$$HCO_3^- + H_2O \Leftrightarrow CO_3^{2-} + H_3O^+ \qquad K_a = 4.8 \times 10^{-11}$$
$$[CO_3^{2-}] = 4.8 \times 10^{-11}\,M$$

Summarizing yields

$$[H_3O^+] = 2.59 \times 10^{-6} \, M \qquad pH = 5.59$$
$$[H_2CO_3] = 1.6 \times 10^{-5} \, M$$
$$[HCO_3^-] = 2.59 \times 10^{-6} \, M$$
$$[CO_3^{2-}] = 4.88 \times 10^{-11} \, M$$

Thus, a pH value of less than 5.6 for a rain or snow sample is due to mineral acids from atmospheric pollution or volcanic emissions. Interaction of less acidic precipitation with soil minerals usually adds alkalinity and raises the pH value, which counteracts the use of carbon dioxide by algae during daylight hours. If the consumption rate of CO_2 is greater than its replacement rate from the atmosphere, as can occur when acid precipitation is input, the dissolved CO_2 concentration in the surface water and groundwater will fall and result in a shift to the left for the corresponding equilibrium reactions:

$$CO_2(aq) + H_2O \Leftrightarrow H_2CO_3$$
$$H_2CO_3 + H_2O \Leftrightarrow HCO_3^- + H_3O^+$$

This will also result in an increase in the pH of the water. As the pH continues to increase, the alkalinity changes chemical species to replace the CO_2 consumed by the algae. Note the equilibrium shifts toward increased CO_2 concentrations, which is illustrated in the following reactions

$$2HCO_3^- \Leftrightarrow CO_3^{2-} + H_2O + CO_2$$
$$CO_3^{2-} + H_2O \Leftrightarrow 2OH^- + CO_2$$

It should be noted that even though we are creating hydroxide alkalinity, the total alkalinity has not changed, merely shifted in chemical form. We define *hydroxide alkalinity* later as alkalinity in excess of a pH value of 10.7. Algae can continue to consume CO_2 until the pH of the water has risen to between 10 and 11, when a growth inhibitory pH is reached and algae consumption of CO_2 is halted. This can result in a diurnal shift in the pH of the photic zone of a water body. In waters containing significant calcium concentrations, the set of reactions above can result in the precipitation of $CaCO_3$ on leaves and twigs in water, and in the long term, can lead to the formation of marl deposits in sediments. Thus, even algae can produce the industrial-sounding "hydroxide alkalinity."

REFERENCES

American Water Works Association, *Standard Methods for the Examination of Water and Wastewater* 18th ed., AWWA, Denver, CO, 1992.

Harris, D. C., *Quantitative Chemical Analysis*, 5th ed., W. H. Freeman, New York, 1998.

Keith, L. H., *Compilation of EPA's Sampling and Analysis Methods*, Lewis Publishers, Chelsea, MI, 1992.

Pohland, F. G. and D. E. Bloodgood, *J. Water Pollut. Control Fed.*, **35**, 11 (1963).

Snoeyink, V. L. and D. Jenkins, *Water Chemistry*, Wiley, New York, 1980.

Sawyer, C. N. and P. L. McCarty, *Chemistry for Environmental Engineering*, 3rd ed., McGraw-Hill, New York, 1978.

Stumm W. and J. J. Morgan, *Aquatic Chemistry*, 3rd ed., Wiley, New York, 1995.

IN THE LABORATORY

To determine the alkalinity, a known volume of water sample is titrated with a standard solution of strong acid to a pH value of approximately 4 or 5. Titrations can be used to distinguish between three types of alkalinity: hydroxide, carbonate, and bicarbonate alkalinity. Carbonate alkalinity is determined by titration of the water sample to the phenolphthalein or metacresol purple indicator endpoint, approximately pH 8.3. Total alkalinity is determined by titration of the water sample to the endpoint of the methyl orange, bromocresol green, or bromocresol green–methyl red indicators, approximately pH 4.5. The difference between the two is the *bicarbonate alkalinity. Hydroxide* (OH^-) *alkalinity* is present if the *carbonate*, or phenolphthalein, *alkalinity* is more than half of the total alkalinity [American Water Works Association (AWWA), 1992]. Thus, the hydroxide alkalinity can be calculated as two times the phenolphthalein alkalinity minus the total alkalinity.

Note that only approximate pH values can be given for the final endpoint, which occurs near a pH value of 4.5. This is because the exact endpoint at the end of the titration, the *equivalence point*, is dependent on the total concentration of carbonate species in solution, while the indicator color change is referred to as the *endpoint*. The endpoint is subject to the pH value only where the indicator changes color and is not influenced by the total alkalinity in solution, whereas the equivalence point is inversely related to alkalinity, with higher total alkalinity corresponding to equivalence at a lower pH value. This can be explained by looking at a pC–pH diagram of the carbonate system. A pC–pH exercise is included in this manual (Chapter 23), and a pC–pH program is included on the accompanying CD-ROM. Figure 21-1 is for a 0.0010 M total carbonate system. The exact equivalence point for the alkalinity titration occurs when the H^+ concentration equals the HCO_3^- concentration. For the 0.001 M carbonate solution (Figure 21-1), this corresponds to the location of the arrow at pH 4.67. As the carbonate concentration increases to 0.10 M (Figure 21-2), the carbonate species lines shift to yield an interception at a pH value of 3.66. This is a significant difference in equivalence points but is not reflected in the indicator endpoint. As a result, the equivalence points described below have been suggested. The following endpoints, corresponding to total alkalinity concentrations, are suggested in AWWA (1992): pH 5.1 for total alkalinities of about 50 mg/L, pH 4.8 for 150 mg/L, and pH 4.5 for 500 mg/L.

Two points should be noted about the titration curve (again, refer to the pC–pH diagrams in Figures 21-1 and 21-2).

1. At pH 10.7, the $[HCO_3^-]$ equals the $[OH^-]$. This is called an equivalence point and is the endpoint of the caustic alkalinity and total acidity titrations. At pH 8.3, the $[H_2CO_3]$ equals the $[CO_3^{2-}]$. This is the endpoint for carbonate alkalinity and CO_2 acidity titrations. In the alkalinity titration virtually all of the CO_3^{2-} has reacted (thus, the term *carbonate alkalinity*) and half of the HCO_3^{2-} has reacted at the endpoint.

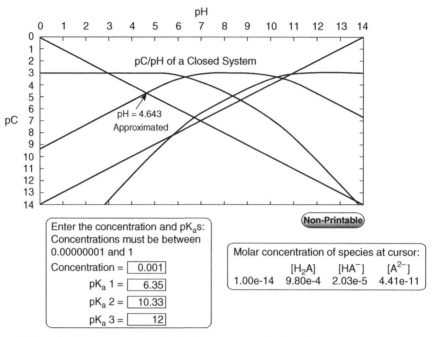

Figure 21-1. pC–pH diagram for a 0.001 M carbonate solution. Refer to and use the pC–pH simulator, which will give color lines on the plot.

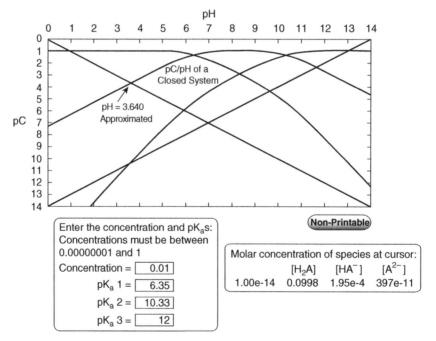

Figure 21-2. pC–pH diagram for a 0.10 M carbonate solution. Refer to and use the pC–pH simulator, which will give color lines on the plot.

2. At pH ~4.5 (dependent on the total alkalinity), the $[H^+]$ equals the $[HCO_3^-]$. This is the endpoint for mineral acidity and total alkalinity titrations.

Safety Precautions

- As in all laboratory exercises, safety glasses must be worn at all times.
- Avoid skin and eye contact with NaOH and HCl solutions. If contact occurs, rinse your hands and/or flush your eyes for several minutes. Seek immediate medical advice for eye contact.
- Use concentrated HCl in the fume hood and avoid breathing its vapor.

Chemicals and Solutions

Sample Handling. Alkalinity is a function of the dissolved CO_2 in solution. Thus, any chemical or physical manipulation of the sample that will affect the CO_2 concentration should be avoided. This includes filtering, diluting, concentrating, or altering the sample in any way. Nor should the sampling temperature be exceeded, as this will cause dissolved CO_2 to be released. Samples containing oil and grease should be avoided. Sampling and storage vessels can be plastic or glass without headspace.

- *Sodium carbonate solution*, ~0.025 *M*. Primary standard grade Na_2CO_3 must be dried for 3 to 4 hours at 250°C and be allowed to cool in a desiccator. Weigh 0.25 g to the nearest 0.001 g and quantitatively transfer all of the solid to a 100-mL volumetric flask. Dilute to the mark with distilled or deionized water. Calculate the exact molarity of the solution in the 100-mL flask.
- *Standardized hydrochloric acid* (about 0.02 *M*). Add 8.3 mL of concentrated (12 *M*) HCl to a 1000-mL volumetric flask and dilute to the mark with deionized or distilled water. This solution has a molarity of approximately 0.10 *M*. Transfer 200 mL of this solution to another 1000-mL volumetric flask to prepare the 0.020 *M* solution. Standardize the dilute HCl solution (about 0.020 *M*) against the Na_2CO_3 primary standard solution. This is done by pipetting 10.00 mL of the ~0.025 *M* Na_2CO_3 solution into a 250-mL Erlenmeyer flask and diluting to about 50 mL with distilled or deionized water. Add 3 to 5 drops of the bromocresol green indicator (more if needed) to the Erlenmeyer flask and titrate with ~0.02 *M* HCl solution. Bromocresol green changes from blue to yellow as it is acidified. The indicator endpoint is intermediate between blue and yellow, and appears as a distinct green color. Determine the molarity of the HCl solution. Remember to wash down any droplets of solution from the walls of the flask.
- *Bromocresol green indicator solution*, about 0.10%, pH 4.5 indicator. Dissolve 0.100 g of the sodium salt into 100 mL of distilled or deionized water. Colors: yellow in acidic solution, blue in basic solution.

- *Phenolphthalein solution*, alcoholic, pH 8.3 indicator. Colors: colorless in acidic solution, red in basic solution.
- *Metacresol purple indicator solution*, pH 8.3 indicator. Dissolve 100 mg of metacresol purple in 100 mL of water. Colors: yellow in acidic solution, purple in basic solution.
- *Mixed bromocresol green–methyl red indicator solution*. You may use either the water- or alcohol-based indicator solution. Water solution: dissolve 100 mg of bromocresol green sodium salt and 20 mg of methyl red sodium salt in 100 mL of distilled or deionized water. Ethyl or isopropyl alcohol solution: dissolve 100 mg of bromocresol green and 20 mg of methyl red in 100 mL of 95% alcohol.

Glassware

- Standard laboratory glassware: 50-mL buret, 250-mL Erlenmeyer flasks, 50-mL beakers, Pasteur pipets

PROCEDURE

Limits of the Method. Typically, 20 mg of $CaCO_3$/L. Lower detection limits can be achieved by using a 10-mL microburet (Keith, 1992)

1. First, an adequate sample volume for titration must be determined. This is accomplished by performing a test titration. Select a volume of your sample, such as 100 mL, and titrate it to estimate the total alkalinity of your sample. For best accuracy, you should use at least 10 mL but not more than 50 mL from a 50-mL buret. Adjust your sample size to meet these criteria.

2. Titrate your sample with standardized 0.02 M HCl solution. Add phenolphthalein or metacresol purple indicator solution and note the color change around a pH value of 8.3. Alternatively, a pH meter can be used to determine the inflection point. This measurement will be a combination of the hydroxide and carbonate alkalinity.

3. Continue the titration to the ~4.5 endpoint by adding bromocresol green or the mixed bromocresol green–methyl red indicator solution. Better results will be obtained by titrating a new sample to the ~4.5 endpoint. This will avoid potential color interferences between the 8.3 and 4.5 pH indicators. Note the color change near a pH value of 4.5. Alternatively, a pH meter can be used to determine the inflection point.

4. Repeat steps 2 and 3 at least three times (excluding the trial titration to determine your sample volume).

5. Calculate the hydroxide, carbonate, bicarbonate, and total alkalinities for your samples. Report your values in mg $CaCO_3$/L. Show all calculations in your notebook.

Waste Disposal

After neutralization, all solutions can be disposed of down the drain with water.

ADVANCED STUDY ASSIGNMENT

1. In your own words, define *alkalinity* and explain why it is important in environmental chemistry.

2. What are the primary chemical species responsible for alkalinity in natural waters?

3. Alkalinity can be expressed in three forms: hydroxide alkalinity, carbonate alkalinity, and total alkalinity. Each of these is determined by titration, but at different pH values. What is the approximate endpoint pH for the carbonate alkalinity titration? What is the approximate endpoint pH for the total alkalinity titration?

4. Why can we give only approximate pH endpoints for each type of alkalinity?

5. To prepare yourself for the laboratory exercise, briefly outline a procedure for titrating a water sample for alkalinity. (List the major steps.)

6. If you titrate 200 mL of a sample with 0.0200 M HCl and the titration takes 25.75 mL of acid to reach the bromocresol green endpoint, what is the total alkalinity of the sample?

7. The atmospheric concentration of CO_2 is predicted to increase up to 750 ppm by the year 2100. What will be the pH of rainwater if it is in equilibrium with an atmosphere containing 500 ppm CO_2?

DATA COLLECTION SHEET

DATA COLLECTION SHEET

DATA COLLECTION SHEET

DATA COLLECTION SHEET

22

DETERMINATION OF HARDNESS IN A WATER SAMPLE

Purpose: To learn the EDTA titration method for determining the hardness of a water sample

BACKGROUND

In the past, *water hardness* was defined as a measure of the capacity of water to precipitate soap. However, current laboratory practices define total hardness as the sum of divalent ion concentrations, especially those of calcium and magnesium, expressed in terms of mg $CaCO_3$/L. There are no known adverse health effects of hard or soft water, but the presence of hard waters results in two economic considerations: (1) hard waters require considerably larger amounts of soap to foam and clean materials, and (2) hard waters readily precipitate carbonates (known as *scale*) in piping systems at high temperatures. Calcium and magnesium carbonates are two of the few common salts whose solubility decreases with increasing temperature. This is due to the removal of dissolved CO_2 as temperature increases. The advent of synthetic detergents has significantly reduced the problems associated with hard water and the "lack of foaming." However, scale formation continues to be a problem.

The source of a water sample usually determines its hardness. For example, surface waters usually contain less hardness than do groundwaters. The hardness of water reflects the geology of its source. A color-coded summary of water hardness in the United States can be found at http://www.usgs.org, and if

Environmental Laboratory Exercises for Instrumental Analysis and Environmental Chemistry
By Frank M. Dunnivant
ISBN 0-471-48856-9   Copyright © 2004 John Wiley & Sons, Inc.

TABLE 22-1. Correlation of Water Hardness Values with Degrees of Hardness

mg CaCO$_3$/L Hardness	Degree of Hardness
0–75	Soft
75–150	Moderately hard
150–300	Hard
>300	Very hard

you view this map you will see that hardness values can range from less than 50 mg/L to over 250 mg/L. Therefore, depending on your water's source, some modifications to the procedure described below may be necessary. Carbonates in surface soils and sediments increase the hardness of surface waters, and subsurface limestone formations also increase the hardness of groundwaters. As indicated in Table 22-1, hardness values can range from a few to hundreds of milligrams of CaCO$_3$ per liter.

The divalent metal cations responsible for hardness can react with soap to form precipitates, or when the appropriate anions are present, to form scale in hot-water pipes. The major hardness-causing cations are calcium and magnesium, although strontium, ferrous iron, and manganese can also contribute. It is common to compare the alkalinity values of a water sample to the hardness values, with both expressed in mg CaCO$_3$/L. When the hardness value is greater than the total alkalinity, the amount of hardness that is equal to the alkalinity is referred to as the *carbonate hardness*. The amount in excess is referred to as the *noncarbonate hardness*. When the hardness is equal to or less than the total alkalinity, all hardness is carbonate hardness and no noncarbonate hardness is present. Common cations and their associated anions are shown in Table 22-2.

THEORY

The method described below relies on the competitive complexation of divalent metal ions by ethylenediaminetetraacetic acid (EDTA) or an indicator. The

TABLE 22-2. Common Cation–Anion Associations Affecting Hardness and Alkalinity

Cations Yielding Hardness	Associated Anions
Ca^{2+}	HCO$_3^-$
Mg^{2+}	SO$_4^{2-}$
Sr^{2+}	Cl$^-$
Fe^{2+}	NO$_3^-$
Mn^{2+}	SiO$_3^{2-}$

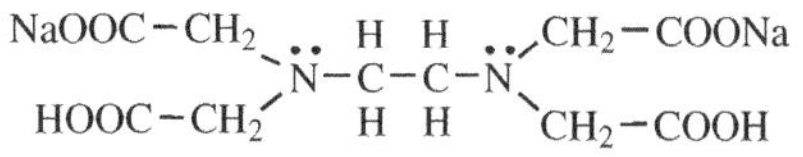

$$\begin{array}{c} NaOOC-CH_2 \\ HOOC-CH_2 \end{array} \!\! \overset{\cdot\cdot}{N}-\overset{H}{\underset{H}{C}}-\overset{H}{\underset{H}{C}}-\overset{\cdot\cdot}{N} \!\! \begin{array}{c} CH_2-COONa \\ CH_2-COOH \end{array}$$

Figure 22-1. Chemical structure for the disodium salt of EDTA.

chemical structure for the disodium salt of EDTA is shown in Figure 22-1. Note the lone pairs of electrons on the two nitrogens. These, combined with the dissociated carboxyl groups, create a 1 : 1 hexadentate complex with each divalent ion in solution. However, the complexation constant is a function of pH (Harris, 1999). Virtually all common divalent ions will be complexed at pH values greater than 10, the pH used in this titration experiment and in most hardness tests. Thus, the value reported for hardness includes all divalent ions in a water sample.

Three indicators are commonly used in EDTA titration, Eriochrome Black T (Erio T), Calcon, and Calmagite. The use of Eriochrome Black T requires that a small amount of Mg^{2+} ion be present at the beginning of the titration. Calmagite is used in this experiment because its endpoint is sharper than that of Eriochrome Black T.

REFERENCES

American Water Works Association, *Standard Methods for the Examination of Water and Wastewater*, 18th ed., AWWA, Denver, Co, 1992.

Harris, D. C., *Quantitative Chemical Analysis*, 5th ed., W.H. Freeman and Company, New York, 1998.

Keith, L. H., *Compilation of EPA's Sampling and Analysis Methods*, Lewis Publishers, Chelsea, MI, 1992.

Sawyer, C. N. and P. L. McCarty, *Chemistry for Environmental Engineering*, 3rd ed., McGraw-Hill, New York, 1978.

Snoeyink, V. L. and D. Jenkins, *Water Chemistry*, Wiley, New York, 1980.

IN THE LABORATORY

Two methods are available for determining the hardness of a water sample. The method described and used here is based on a titration method using a chelating agent. The basis for this technique is that at specific pH values, EDTA binds with divalent cations to form a strong complex. Thus, by titrating a sample of known volume with a standardized (known) solution of EDTA, you can measure the amount of divalent metals in solution. The endpoint of the titration is observed using a colorimetric indicator, in our case Calmagite. When a small amount of indicator is added to a solution containing hardness (at pH 10.0), it combines with a few of the hardness ions and forms a weak wine red complex. During the titration, EDTA complexes more and more of the hardness ions until it has complexed all of the free ions and "outcompetes" the weaker indicator complex for hardness ions. At this point, the indicator returns to its uncomplexed color (blue for Calmagite), indicating the endpoint of the titration, where only EDTA-complexed hardness ions are present.

Safety Precautions

- As in all laboratory exercises, safety glasses must be worn at all times.
- Avoid skin and eye contact with pH 10 buffer. In case of skin contact, rinse the area for several minutes. For eye contact, flush eyes with water and seek immediate medical advice.

Chemicals and Solutions

Sample Handling. Plastic or glass sample containers can be used. A minimum of 100 mL is needed, but for replicate analysis of low-hardness water, 1 L of sample is suggested. If you are titrating the sample on the day of collection, no preservation is needed. If longer holding times are anticipated, the sample can be preserved by adding nitric or sulfuric acid to a pH value of less than 2.0. Note that this acidic pH level must be adjusted to above a pH value of 10 before the titration.

- *pH 10 buffer.* In a 250-mL volumetric flask, add 140 mL of a 28% by weight NH_3 solution to 17.5 g of NH_4Cl and dilute to the mark.
- *Calmagite* [1-(1-hydroxy-4-methyl-2-phenylazo)-2-naphthol-4-sulfonic acid]. Dissolve 0.10 g of Calmagite in 100 mL of distilled or deionized water. Use about 1 mL per 50-mL sample to be titrated.
- *Analytical reagent-grade Na_2EDTA* (FW 372.25). Dry at 80°C for 1 hour and cool in a desiccator. Accurately weigh 3.723 g (or a mass accurate to 0.001 g), dissolve in 500 mL of deionized water with heating, cool to room

temperature, quantitatively transfer to a 1-L volumetric flask, and fill to the mark. Since EDTA will extract hardness-producing cations out of most glass containers, store the EDTA solution in a plastic container. This procedure produces a 0.0100 M solution.

Glassware

- Standard laboratory glassware: 50-mL buret, 250-mL Erlenmeyer flasks, 50-mL beakers, Pasteur pipets

PROCEDURE

Limits of the Method. Detection limits depend on the volume of sample titrated.

1. Pipet an aliquot of your sample into a 250-mL Erlenmeyer flask. The initial titration will only be a trial and you will probably need to adjust your sample volume to obtain the maximum precision from your pipetting technique (use more than 10 mL but less than 50 mL). Increase or decrease your sample size as needed.
2. Add 3 mL of the pH 10 buffer solution and about 1 mL of the Calmagite indicator. Check to ensure that the pH of your sample is at or above pH 10. Add additional buffer solution if needed.
3. Titrate with EDTA solution and note the color change as you reach the endpoint. Continue adding EDTA until you obtain a stable blue color with no reddish tinge (incandescent light can produce a reddish tinge at and past the endpoint).
4. Repeat until you have at least three titrations that are in close agreement.
5. Calculate the hardness for each of your samples. Express your results in mg $CaCO_3$/L. If you made the EDTA solution exactly according to the procedure, 1.00 mL of EDTA solution is equal to 1.00 mg $CaCO_3$/L. Confirm this through calculations.

Waste Disposal

After neutralization, all solutions can be disposed of down the drain with rinsing.

ADVANCED STUDY ASSIGNMENT

1. In your own words, define *hardness*.
2. What are the primary cations typically responsible for hardness?
3. In what unit of measure is hardness usually expressed?
4. What is meant by *carbonate* and *noncarbonate hardness*?
5. What is the color change for the Calmagite indicator?
6. Briefly outline a procedure for titrating a water sample for hardness. (List the major steps.)
7. If you titrate 50.0 mL of a sample with 0.100 M EDTA and the titration takes 25.75 mL of EDTA to reach the endpoint, what is the hardness of the sample in mg $CaCO_3/L$?

DATA COLLECTION SHEET

DATA COLLECTION SHEET

DATA COLLECTION SHEET

DATA COLLECTION SHEET

DATA COLLECTION SHEET

PART 7

FATE AND TRANSPORT CALCULATIONS

23

pC–pH DIAGRAMS: EQUILIBRIUM DIAGRAMS FOR WEAK ACID AND BASE SYSTEMS

Purpose: To learn to plot and interpret pC–pH diagrams manually

BACKGROUND

The concentration of a weak acid or base in a solution (e.g., H_2CO_3, HCO_3^-, or CO_3^{2-}) can be calculated using simple equilibrium expressions at any given pH. In some cases it is useful to look at the equilibrium distribution of each of the protonated and nonprotonated species in solution at the same time. A pC–pH diagram such as those shown in Figures 23-1 and 23-2 is an excellent tool for viewing these concentrations simultaneously. As the name implies, the concentrations of all chemical species (including the hydronium ion) are expressed as the negative log of concentration (for the hydronium ion, the pH). To construct a pC–pH diagram, the total concentration of the acid or base is needed along with the corresponding equilibrium equations and constants (K).

CLOSED SYSTEMS

All pC–pH diagrams have two lines in common, the line describing the concentration of hydroxide (OH^-) as a function of pH and the line describing

Environmental Laboratory Exercises for Instrumental Analysis and Environmental Chemistry
By Frank M. Dunnivant
ISBN 0-471-48856-9   Copyright © 2004 John Wiley & Sons, Inc.

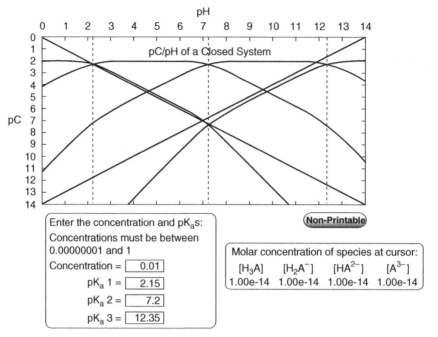

Figure 23-1. pC–pH diagram for a triprotic system.

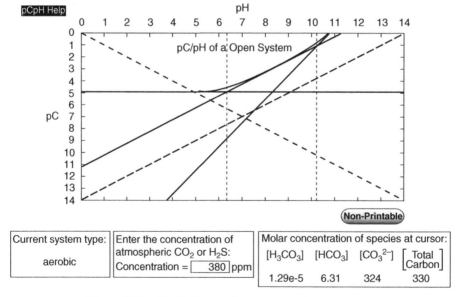

Figure 23-2. pC–pH diagram for an open carbonate system.

the concentration of hydronium ion (H^+) as a function of pH. These are based on the equilibrium relation

$$H_2O \Leftrightarrow H^+ + OH^-$$

where

$$K_w = [H^+][OH^-] = 1 \times 10^{-14}$$

By rearranging and taking the negative log of each side, we obtain

$$-\log K_w = -\log[H^+] - \log[OH^-]$$

$$pOH = 14 - pH$$

The slope of the diagonal line representing the change in $[OH^-]$ and $[H^+]$ is

$$\frac{\Delta pOH}{\Delta pH} = -1$$

and when pH equals 0, the pOH equals 14. This results in a line from (pH 0.0, pC 14.0) to (pH 14.0, pC 0.0). Similarly, a line can be drawn representing the hydronium ion concentration as a function of pH. By definition

$$-\log[H^+] = pH$$

Therefore,

$$\frac{\Delta(-\log[H^+])}{pH} = 1$$

When the pH equals 0, $-\log[H^+]$ equals 0. This results in a line from (pH 0.0, pC 0.0) to (pH 14.0, pC 14.0).

The next line (or set of lines) normally drawn on a pC–pH diagram is the one representing the total concentration of acid or base, C_T. When pC–pH diagrams are drawn by hand, C_T is drawn as a straight horizontal line starting at pC_T on the y axis. This line is actually a combination of two or more lines, depending on the number of protons present in the acid. At the first pK_a (the negative log of the K_a), two lines intersect, one with a negative whole-number slope and one with a positive whole-number slope. For diprotic and triprotic systems, as one species line crosses a second line (or pK_a), the slope of the line shifts from -1 or $+1$ to -2 or $+2$, respectively. These lines represent the concentration of each chemical species. Three cases are given below: a triprotic system (the phosphate system), a diprotic system (the carbonate system), and a monoprotic system (a generic system).

For a *triprotic system*, the lines for each individual chemical species can be represented by

$$H_3A \Leftrightarrow H_2A^- + H^+ \quad \text{where } K_1 = \frac{[H_2A^-][H^+]}{[H_3A]}$$

$$H_2A^- \Leftrightarrow HA^{2-} + H^+ \quad \text{where } K_2 = \frac{[HA^{2-}][H^+]}{[H_2A^-]}$$

$$HA^{2-} \Leftrightarrow A^{3-} + H^+ \quad \text{where } K_3 = \frac{[A^{3-}][H^+]}{[HA^{2-}]}$$

The total concentration of the acid or base, C_T, is a sum of all protonated and nonprotonated species, such that

$$C_T = H_3A + H_2A^- + HA^{2-} + A^{3-}$$

When the equilibrium expressions above and the C_T equation are combined and solved for the concentrations of H_3A, H_2A^-, HA^{2-}, and A^{3-} in terms of C_T, $[H^+]$, and the equilibrium constants, four equations are obtained:

$$[H_3A] = C_T \frac{1}{1 + (K_1/[H^+]) + (K_1K_2/[H^+]^2) + (K_1K_2K_3/[H^+]^3)}$$

$$[H_2A^-] = C_T \frac{1}{([H^+]/K_1) + 1 + (K_2/[H^+]) + (K_2K_3/[H^+]^2)}$$

$$[HA^{2-}] = C_T \frac{1}{([H^+]^2/K_1K_2) + ([H^+]/K_2) + 1 + (K_3/[H^+])}$$

$$[A^{3-}] = C_T \frac{1}{([H^+]^3/K_1K_2K_3) + ([H^+]^2/K_2K_3) + ([H^+]/K_3) + 1}$$

If a pH-dependent constant α_H is defined as

$$\alpha_H = \frac{[H^+]^3}{K_1K_2K_3} + \frac{[H^+]^2}{K_2K_3} + \frac{[H^+]}{K_3} + 1$$

the previous equations can be simplified to:

$$[H_3A] = \frac{C_T[H^+]^3}{K_1K_2K_3\alpha_H}$$

$$[H_2A^-] = \frac{C_T[H^+]^2}{K_2K_3\alpha_H}$$

$$[HA^{2-}] = \frac{C_T[H^+]}{K_3\alpha_H}$$

$$[A^{3-}] = \frac{C_T}{\alpha_H}$$

Several important points about the pC–pH diagram should be noted. Figure 23-1 was made for a 0.001 M phosphate system. Note the lines representing the hydrogen and hydroxide ion concentrations, which have a slope of $+1$ and -1, respectively. The system points, defined as vertical lines at each pK_a value, represent the pH where each chemical species adjacent to these lines is at equal concentration. Note that this condition is met at each equilibrium constant (K_a), or using the negative log scale, at each pK_a value. For the phosphate system these values are $10^{-2.1}$ for K_1, $10^{-7.2}$ for K_2, and $10^{-12.3}$ for K_3. Note that at pH values less than pK_1, the system is dominated by the H_3PO_4 species; at pH values between pK_1 and pK_2, the system is dominated by the $H_2PO_4^-$ ion; at pH values between pK_2 and pK_3, the system is dominated by the HPO_4^{2-} ion; and at pH values above pK_3, the system is dominated by the PO_4^{3-} ion. Also note that as the pH increases, each time the line describing the concentration of a species approaches a pK_a value and crosses another pK_a (system point), the slope of the line decreases by a whole-number value.

For a *diprotic system*, the equilibrium equations for H_2A, HA^-, and A^{2-} are

$$H_2A \Leftrightarrow HA^- + H^+ \qquad \text{where } K_1 = \frac{[HA^-][H^+]}{[H_2A]}$$

$$HA^- \Leftrightarrow A^{2-} + H^+ \qquad \text{where } K_2 = \frac{[A^{2-}][H^+]}{[HA^-]}$$

When these equations are combined with the mass balance equation,

$$C_T = H_2A + HA^- + A^{2-}$$

and solved for H_2A, HA^-, and A^{2-} in terms of C_T, $[H^+]$, and the equilibrium constants, three equations are obtained:

$$[H_2A] = C_T \frac{1}{1 + (K_1/[H^+]) + (K_1 K_2/[H^+]^2)}$$

$$[HA^-] = C_T \frac{1}{([H^+]/K_1) + 1 + (K_2/[H^+])}$$

$$[A^{2-}] = C_T \frac{1}{([H^+]^2/K_2 K_2) + ([H^+]/K_2) + 1}$$

If a pH-dependent constant α_H is defined as

$$\alpha_H = \frac{[H^+]^2}{K_1 K_2} + \frac{[H^+]}{K_2} + 1$$

the previous equations for the diprotic system can be simplified to

$$[H_2A] = \frac{C_T[H^+]^2}{K_1 K_2 \alpha_H}$$

$$[HA^-] = \frac{C_T[H^+]}{K_2 \alpha_H}$$

$$[A^{2-}] = \frac{C_T}{\alpha_H}$$

For a *monoprotic system*, the governing equations are considerably simpler, since only pK_a is involved,

$$[HA] = \frac{C_T[H^+]}{[H^+] + K_a} \quad \text{and} \quad [A^-] = \frac{C_T K_a}{[H^+] + K_a}$$

The utility of a pC–pH diagram is that all of the ion concentrations can be estimated at the same time for any given pH value. This computer simulation in the pC–pH computer package included with your manual allows the user to select an acid system, enter the pK_a values, and draw the pC–pH diagram. After the diagram is drawn, the user can point the cursor at a given pH, and the concentrations of each ion will be given. Additional discussions of pC–pH diagrams can be found in Langmuir (1997) and Snoeyink and Jenkins (1980).

OPEN SYSTEMS

The pC–pH diagrams for open systems are similar to those described for closed systems. The primary difference is that in an open system a component of the system exists as a gas and the system is open to the atmosphere. In other words, the system can exchange matter and energy with the atmosphere. The most important environmental example of such a system comprises carbon dioxide (CO_2), carbonic acid (H_2CO_3), bicarbonate ion (HCO_3^-), and carbonate ion (CO_3^{2-}) in lakes, rivers, and oceans.

The reactions occurring in this system are

$$CO_2 + H_2O \leftrightarrow H_2CO_3$$

$$H_2CO_3 \leftrightarrow HCO_3^- + H^+$$

$$HCO_3^- \leftrightarrow CO_3^{2-} + H^+$$

$$H_2O \leftrightarrow H^+ + OH^-$$

The equilibrium relationships for this system are

$$K_w = [H^+][OH^-] = 10^{-14}$$

$$K_{CO_2} = \frac{[H_2CO_3]}{P_{CO_2}} = 10^{-1.47}$$

$$K_1 = \frac{[H^+][HCO_3^-]}{[H_2CO_3]} = 10^{-6.35}$$

$$K_2 = \frac{[H^+][CO_3^{2-}]}{[HCO_3^-]} = 10^{-10.33}$$

where P_{CO_2} is the partial pressure of CO_2 in the atmosphere.

Open system pC–pH diagrams contain lines describing the concentration of hydroxide (OH^-) and hydronium ion (H^+) identical to those for closed systems. However, because open systems can exchange matter with the atmosphere, the total inorganic carbon concentration is not constant as it is for a closed system, but varies as a function of pH. Still, the total inorganic carbon concentration is the sum of all inorganic carbon species, as it was for closed systems. In this case,

$$C_T = [H_2CO_3] + [HCO_3^-] + [CO_3^{2-}]$$

The concentration of H_2CO_3, HCO_3^-, and CO_3^{2-} as a function of pH and P_{CO_2} can be calculated from the equilibrium relationships given previously. The equations for these lines are

$$[H_2CO_3] = (K_{CO_2})(P_{CO_2}) = (P_{CO_2})(10^{-1.47})$$

$$-\log[H_2CO_3] = -\log(P_{CO_2}) + 1.47$$

$$[HCO_3^-] = \frac{(K_1)(P_{CO_2})(10^{-1.47})}{H^+} = \frac{(10^{-6.35})(P_{CO_2})(10^{-1.47})}{H^+}$$

$$-\log[HCO_3^-] = -\log(P_{CO_2}) + 7.82 - pH$$

$$[CO_3^{2-}] = \frac{(K_2)(10^{-6.35})(P_{CO_2})(10^{-1.47})}{(H^+)^2} = \frac{(10^{-10.33})(10^{-6.35})(P_{CO_2})(10^{-1.47})}{(H^+)^2}$$

$$-\log[CO_3^{2-}] = -\log(P_{CO_2}) + 18.15 - 2pH$$

As mentioned previously and demonstrated by the equations above, the concentrations of H_2CO_3, HCO_3^-, and CO_3^{2-} vary as a function of both pH and P_{CO_2}. This means that as P_{CO_2} has varied naturally over the years during ice ages and periods of warming, the concentration of H_2CO_3, HCO_3^-, and CO_3^{2-} in surface waters has changed. It also means that P_{CO_2} changes caused by global warming will alter the surface water concentrations of these species.

A pC–pH diagram for the open carbonate system (380 ppm CO_2 in the atmosphere) is shown in Figure 23-2.

REFERENCES

Langmuir, D., *Aqueous Environmental Geochemistry*, Prentice Hall, Upper Saddle River, NJ, 1997.

Snoeyink, V. L. and D. Jenkins, *Water Chemistry*, Wiley, New York, 1980.

ASSIGNMENT

Insert the CD-ROM or install the pC–pH module on your computer (the pC–pH simulator is included with your lab manual). After you have installed pC–pH, if it does not start automatically, open it. A sample data set will load automatically. Work through the example problem, referring to the background information given earlier and the explanation of the example problem (included in the pC–pH module) as needed.

1. Why is the slope of the $[OH^-]$ and $[H^+]$ lines equal to 1.00 and -1.00, respectively?
2. Why does the slope of each carbon species shift by one whole-number value when the line crosses a second pK_a value?
3. Using graph paper, draw a pC–pH diagram manually for a closed carbonate system (total carbonate concentration of 0.0500 M). What is the dominant carbon species at pH 4.0, 8.0, and 11.0? Calculate the exact molar concentration of each chemical species at pH 8.00.
4. Using graph paper, draw a pC–pH diagram manually for an open carbonate system (total atmospheric CO_2 concentration of 450 ppm). What is the dominant carbon species at pH 8.00? Calculate the exact molar concentration of each chemical species at pH 9.50.

To Print a Graph from Fate

For a PC

- Select the printable version of your plot (lower right portion of the screen).
- Place the cursor over the plot at the desired x and y coordinates.
- Hold the alt key down and press print screen.
- Open your print or photoshop program.
- Paste the Fate graph in your program by holding down the control key and press the letter v.
- Save or print the file as usual.

For a Mac

- Select the printable version of your plot.
- Hold down the shift and open apple key and press the number 4. This will place a cross-hair symbol on your screen. Position the cross-hair symbol in the upper right corner of your plot, click the cursor and drag the cross-hair symbol over the area to be printed or saved, release the cursor when you have selected the complete image. A file will appear on your desktop as picture 1.
- Open the file with preview or any image processing file and print it as usual.

24

FATE AND TRANSPORT OF POLLUTANTS IN RIVERS AND STREAMS

Purpose: To learn two basic models for predicting the fate and transport of pollutants in river systems

BACKGROUND

The close proximity to natural waterways of chemical factories, railways, and highways frequently leads to unintentional releases of hazardous chemicals into these systems. Once hazardous chemicals are in the aquatic system, they can have a number of detrimental effects for considerable distances downstream from their source. This exercise allows the user to predict the concentration of a pollutant downstream of an instantaneous release. Examples of instantaneous releases can be as simple as small discrete releases such as dropping a liter of antifreeze off a bridge, or they can be more complex, such as a transportation accident that results in the release of acetone from a tanker car. Continuous (step) releases usually involve steady input from an industrial process, drainage from nonpoint sources, or leachate from a landfill. Once released to the system, the model assumes that the pollutant and stream water are completely mixed (i.e., there is no cross-sectional concentration gradient in the stream channel). This is a reasonably good assumption for most systems. The model used here accounts for longitudinal dispersion (spreading in the direction of stream flow), advection (transport in the direction of stream flow at the flow rate of the water), and a first-order removal term (biodegradation or radioactive decay).

Environmental Laboratory Exercises for Instrumental Analysis and Environmental Chemistry
By Frank M. Dunnivant
ISBN 0-471-48856-9   Copyright © 2004 John Wiley & Sons, Inc.

CONCEPTUAL DEVELOPMENT OF GOVERNING FATE AND TRANSPORT EQUATION

Instantaneous Pollutant Input

Before we show the mathematical development of the governing equation, we present a conceptual approach that shows how each part of the equation relates to a physical model of a polluted river (illustrated in Figure 24-1). The governing equation for the instantaneous model and a typical concentration–time profile for this equation are shown in the upper right-hand corner of the figure. The river is shown flowing from the upper left-hand corner to the lower right-hand corner. The instantaneous source (W in mass units) is shown upstream in the river as an irregular by shaped object. This represents a one-time sporadic input of pollutant, such as a barrel of waste falling in the river or a shipping accident. Upon entry to the river, the pollutant is mixed rapidly and evenly across the cross section of the stream. Next, the velocity gradients (v) and flow rate (Q) are shown. As a plume of pollution is transported down a stream, additional mixing occurs and the length of the pollutant plume increases. We account for this mixing and dilution of the pollutant concentration with E, the longitudinal dispersion coefficient (m^2/s). This is easy but costly to measure in a stream, but we can estimate it accurately by knowing the slope of the stream channel (the decrease in elevation with distance from the pollutant input point). Next, we are concerned with any first-order removal of pollutant from the stream and include microbial and chemical degradations, volatilization, and sorption to river sediments. This accounts for all of the major processes in the real world and all of the terms shown in the governing equation. A typical concentration–distance profile for an instantaneous input is shown in the upper right-hand corner of Figure 24-1, below the instantaneous input model equation.

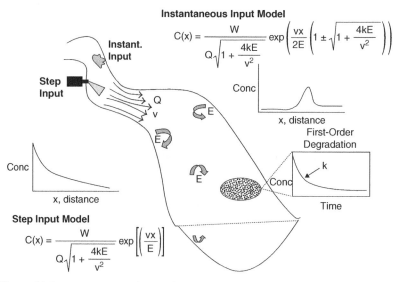

Figure 24-1. Transport equations and conceptualization of a polluted stream system.

Step Pollutant Input

The conceptual approach for the step input of pollutant to a river is very similar to that of an instantaneous input. All of the terms described above are applicable to the step model. However, here the pollution enters the rivers at a constant rate. For example, industries located along the river have permits for federal and state agencies to emit a small amount of waste to the stream. Most industries operate 24 hours a day and 365 days a year, and their process (waste) does not usually change drastically. So we can model the introduction of waste to the river as a constant input. The resulting concentration of a pollutant downstream is a function of mixing and dilution by the river water (described by E) and any degradation or removal that may occur (described by k). A typical concentration–distance profile for a step input is shown in the lower left-hand corner of Figure 24-1 above the step input model label.

Mathematical Approach to a Lake System

The governing equation is obtained initially by setting up a mass balance on a cross section of the stream channel, as described by Metcalf & Eddy (1972). When the dispersion term (E) given above is included in a cross-sectional mass balance of the stream channel, each term can be described as follows

$$\text{Inflow}: \quad QC\,\Delta t - EA\frac{\partial C}{\partial x}\Delta t$$

$$\text{Outflow}: \quad Q\left(C + \frac{\partial C}{\partial x}\Delta x\right)\Delta t - EA\left(\frac{\partial C}{\partial x} + \frac{\partial^2 C}{\partial x^2}\Delta x\right)\Delta t$$

$$\text{Sinks}: \quad vkC\,\Delta t$$

where Q is the volumetric flow rate (m^3/s), C the concentration (mg/m^3), E the longitudinal dispersion coefficient (m^2/s), A the cross-sectional area (m^2), x the distance downstream from point source (m), and v the average water velocity (m/s).

The two longitudinal dispersion terms in these equations,

$$EA\frac{\partial C}{\partial x}\Delta t \quad \text{and} \quad EA\left(\frac{\partial C}{\partial x} + \frac{\partial^2 C}{\partial x^2}\Delta x\right)\Delta t$$

were derived from the equation

$$\frac{\partial M}{\partial t} = -EA\frac{\partial C}{\partial x}$$

where $\partial M/\partial t$ is the mass flow, $\partial C/\partial x$ the concentration gradient, A the cross-sectional area, and E the coefficient of turbulent mixing.

From this equation it can be seen that whenever a concentration gradient exists in the direction of flow ($\partial C/\partial x$), a flow of mass ($\partial M/\partial t$) occurs in a manner to reduce the concentration gradient. For this equation it is assumed that the flow rate is proportional to the concentration gradient and the cross-sectional area over which this gradient occurs. The proportionality constant, E, is commonly called the *coefficient of eddy diffusion* or *turbulent mixing*. Thus, the driving force behind this reduction in concentration is the turbulent mixing in the system, characterized by E and the concentration gradient.

The inflow, outflow, and sink equation given earlier can be combined to yield the pollutant concentration at a given cross section as a function of time. This combination of terms is generally referred to as the *general transport equation* and can be expressed as

$$\text{accumulation} = \text{inputs} - \text{outputs} + \text{sources} - \text{removal}$$

Instantaneous Pollutant Input Model

Combining the inflow, outflow, instantaneous source, and sink terms into the mass balance expression and integrating for the equilibrium case where $\partial C/\partial t = 0$ results in the following governing equation for the transport of an instantaneous input to a stream system:

$$C_{(x,t)} = \frac{M_0}{Wd\sqrt{4\pi Et}} \exp\left[\frac{-(x - vt)^2}{4Et} - kt\right] \qquad (24\text{-}1)$$

where $C(x)$ = pollutant concentration (mg/L or µCi/L for radioactive
          compounds) at distance x and time t
M_0 = mass of pollutant released (mg or µCi)
W = average width of the stream (m)
d = average depth of the stream (m)
E = longitudinal dispersion coefficient (m²/s)
t = time (s)
x = d/t; distance downstream from input (m)
v = average water velocity (m/s)
k = first-order decay or degradation rate constant (s⁻¹)

Note that exp represents e (the base of the natural logarithm).

When there is no (or negligible) degradation of the pollutant, k is set to zero (or a very small number in Fate). The longitudinal dispersion coefficient, E, is characteristic of the stream, or more specifically, the section of the stream that is being modeled. Values of E can be determined experimentally by adding a known mass of tracer to the stream and measuring the tracer concentration at various points as a function of time. Equation (24-1) is then fitted to the data at each sampling point and a value for E is estimated. Unfortunately, this experimental approach is very time and cost intensive, and is rarely used. One common approach for estimating E values is given by Fischer et al. (1979):

$$E = 0.011 \frac{v^2 w^2}{du} \quad \text{and} \quad u = \sqrt{gds}$$

where v is the average water velocity (m/s), w the average stream width (m), d the average stream depth (m), $g = 9.81$ m/s^2 (the acceleration due to gravity), and s the slope of the streambed (unitless).

From these equations it can be seen that the downstream concentration of a pollutant (in the absence of degradation) is largely a function of the longitudinal dispersion, which, in turn, is determined by the mixing in the system and the slope of the streambed.

Step Pollutant Input Model

Combining the inflow, outflow, step source, and sink terms into the mass balance expression and integrating for the equilibrium case where $\partial C/\partial t = 0$ results in the following governing equation for the transport of a step input to a stream system:

$$C(x) = \frac{W}{Q\sqrt{1 + 4kE/v^2}} \exp\left[\frac{vx}{2E}\left(1 \pm \sqrt{1 + \frac{4kE}{v^2}}\right)\right]$$

where $C(x) =$ pollutant concentration (mg/L or µCi/L for radioactive compounds) at distance x and time t
   $W =$ rate of continuous discharge of the waste (kg/s or Ci/s)
   $Q =$ stream flow rate (m^3/s)
   $E =$ longitudinal dispersion coefficient (m^2/s)
   $x =$ distance downstream from input (m)
   $v =$ average water velocity (m/s)
   $k =$ first-order decay or degradation rate constant (s^{-1})

The positive root of the equation refers to the upstream direction $(-x)$, and the negative root (what we use in Fate) refers to the downstream direction $(+x)$.

When there is no (or negligible) degradation of the pollutant, k is set to zero (or a very small number in Fate). The longitudinal dispersion coefficient, E, is characteristic of the stream, or more specifically, the section of the stream that is being modeled. Under these conditions the governing equation reduces to

$$C(x) = \frac{W}{Q\sqrt{1 + 4kE/v^2}} \exp\left[\left(\frac{vx}{E}\right)\right]$$

As in the instantaneous input model, values of E are estimated using the approach outlined by Fischer et al. (1979).

From these equations it can be seen that the downstream concentration of a pollutant (in the absence of degradation) is largely a function of the longitudinal dispersion, which, in turn, is determined by the mixing in the system and the slope of the streambed.

REFERENCES

Fischer, H. B., E. J. List, R. C. Y. Koh, I. Imberger, and N. H. Brooks, *Mixing in Inland and Coastal Waters*, Academic Press, New York, 1979.

Metcalf & Eddy, Inc., *Wastewater Engineering: Collection, Treatment, Disposal*, McGraw Hill, New York, 1972.

ASSIGNMENT

Install Fate on your computer (Fate is included with your lab manual). Open the program and select the river step or pulse module. A sample data set will load automatically. Work through the example problem, referring to the background information above and the explanation of the example problem (included in Fate) as needed.

1. Select a pollutant and conduct the simulations described below for a step and instantaneous pollution scenario. In selecting your pollutant and input conditions, you must use a mass that will be soluble or miscible with water. An important assumption in the governing equation for all fate and transport models is that no pure solid or pure nonmiscible liquid phase of the pollutant is present.

2. Construct a pollution scenario for your simulations. This will require you to input data on a specific river, such as flow rates, background pollutant concentrations, and any pollutant decay rates (most are given in the table of first-order decay rates included in Fate). The U.S. Geological Survey maintains a Web site of stream flow rates in the United States. These can be accessed at `http://www.usgs.org`.

3. Perform a simulation using your basic input data, and evaluate the effluent pollutant concentration for the step and pulse pollution scenarios. Next, perform a sensitivity test by selecting and varying several input variables, such as mass loading, flow rate (to reflect an unusually wet or dry season), and first-order decay rate (those given in the table are only estimates; the actual value can depend on factors such as volatilization, the present of different bacterial communities, temperature, chemical degradations, photochemical degradations, etc.).

4. Write a three- to five-page paper discussing the results of your simulations. Include tables of data and/or printouts of figures from Fate. A copy of your report should be included in your lab manual.

To Print a Graph from Fate

For a PC

- Select the printable version of your plot (lower right portion of the screen).
- Place the cursor over the plot at the desired x and y coordinates.
- Hold the alt key down and press print screen.
- Open your print or photoshop program.
- Paste the Fate graph in your program by holding down the control key and press the letter v.
- Save or print the file as usual.

For a Mac

- Select the printable version of your plot.
- Hold down the shift and open apple key and press the number 4. This will place a cross-hair symbol on your screen. Position the cross-hair symbol in the upper right corner of your plot, click the cursor and drag the cross-hair symbol over the area to be printed or saved, release the cursor when you have selected the complete image. A file will appear on your desktop as picture 1.
- Open the file with preview or any image processing file and print it as usual.

25

FATE AND TRANSPORT OF POLLUTANTS IN LAKE SYSTEMS

Purpose: To learn two basic models for predicting the fate and transport of pollutants in lake systems

BACKGROUND

Lakes and human-made reservoirs serve as valuable drinking water resources. Although many small lakes remain pristine, most human-made lakes suffer from overdevelopment, and large lakes are subject to contamination from local industrial sources and shipping accidents. Regardless of the size of the lake, most introductory modeling efforts simplify the governing equations by assuming that the lake is completely mixed immediately after the addition of a contaminant. It is also assumed that the volume of the lake does not change over the time interval of study, so that the volume of water entering the lake is equal to the volume of water exiting the lake, usually in the form of a stream.

CONCEPTUAL DEVELOPMENT OF GOVERNING FATE AND TRANSPORT EQUATION

Instantaneous Pollutant Input

Before we show the mathematical development of the governing equation, we present a conceptual approach that shows how each part of the equation relates to

Environmental Laboratory Exercises for Instrumental Analysis and Environmental Chemistry
By Frank M. Dunnivant
ISBN 0-471-48856-9   Copyright © 2004 John Wiley & Sons, Inc.

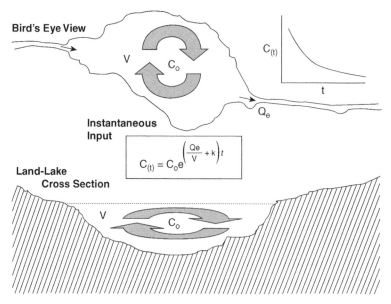

Figure 25-1. Pollutant concentrations in a lake following an instantaneous input.

a physical model of the lake (Figure 25-1). Two views of the lake are shown in this figure. The upper figure shows a bird's-eye view of the lake, with the water entering the lake on the left and exiting on the right. The governing equation is shown in the center of the figure. The concentration of pollutant in the exiting water is shown in the upper right-hand corner of Figure 25-1 as a function of time elapsed since input. The lower figure shows a cross section of the lake.

First we assume that the input of pollutant is evenly distributed over the entire lake and that the lake is completely mixed. Thus, the total mass of pollutant added to the lake is divided by the volume (V) of the lake to yield the initial pollutant concentration, C_0. Next, we look at how pollution is removed from the lake. Our model assumes that there are two ways of removing pollution from the lake: degradation (microbial or chemical) or other loss processes (such as sorption and volatilization) described by the first-order rate constant (k) in the governing equation, and natural removal out of the lake with the river water (represented by Q_e). Since the lake is completely mixed and the pollutant concentration is equal everywhere in the lake, the concentration of pollutant in the exiting river is the same as the concentration in the lake. This concentration is represented by C_t in the governing equation and is the concentration at a specific time after the addition of pollutant to the lake. As time passes (t increases) the concentration of pollutant in the lake and in the exiting water can be calculated using the equation for instantaneous pollutant input. This accounts for all the terms in the governing equation. A more mathematical approach to our modeling effort is described later in this section.

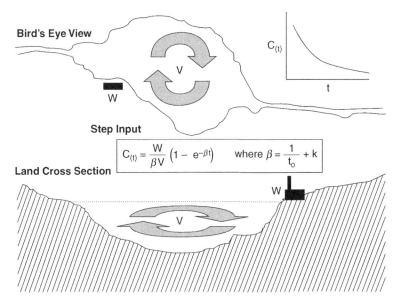

Figure 25-2. Pollutant concentration in a lake undergoing step input.

Step Pollutant Input

The conceptual approach for a step input of pollutant to a lake is similar to that of an instantaneous input. First, the lake water and the pollutant are mixed completely and evenly. However, in the step input, the pollutant is emitted from a point source such as a chemical plant, represented by a W in Figure 25-2. The units of W are mass per time, and this mass is divided by the volume (V) of the lake to yield a concentration (mass/volume). As in the instantaneous example, we treat microbial and chemical degradation as well as volatilization and adsorption reactions as first-order processes represented by k in the equation. Finally, we need to know the residence time of water in the lake. This is calculated by dividing the volume of the lake (V) by the volumetric flow rate of water out of the lake (Q_e), which yields t_0 (the time an average water molecule spends in the lake). Using this approach and the governing equation shown in Figure 25-2, we can calculate the pollutant concentration as a function of time. A typical plot of this type is shown in the upper right-hand portion of Figure 25-2.

Mathematical Approach to a Lake System

The first step in developing the governing equations for the fate of a pollutant in a lake system is to set up a mass balance on the system. First, quantify all of the mass inputs of pollutant to the system. This can be expressed as

$$W = Q_w C_w + Q_i C_i + Q_{\text{trib}} C_{\text{trib}} + PA_s C_p + VC_s \qquad (25\text{-}1)$$

where W = mass input of pollutant rate per unit time (kg/time)
        Q_w = inflow rate of the wastewater (m³/time)
        C_w = pollutant concentration in the wastewater (kg/m³)
        Q_i = inflow rate of the main river (m³/time)
        C_i = pollutant concentration in the main inlet river (kg/m³)
        Q_{trib} = net inflow rate from all other tributaries (m³/time)
        C_{trib} = net pollutant concentration in the tributaries (kg/m³)
        P = annual precipitation (m/time)
        A_s = mean lake surface area (m²)
        C_p = net pollutant concentration in precipitation (kg/m³)
        V = average lake volume (m³)
        C_s = average pollutant release from suspended lake sediments
             (kg/m³ · time)

In most situations, the mass inputs from the smaller tributaries and precipitation are minor compared to the major input source, and these terms are ignored. We will simplify the mass input expression further here by assuming that the contribution from contaminated sediments is negligible, but this is not always the case. These assumptions simplify the input expression to

$$W = Q_w C_w + Q_i C_i \qquad (25\text{-}2)$$

Next, we set up a mass balance for the pollutant across the entire system,

change in mass =   inflow   − outflow + sources −   sinks
$$V\,dC = (Q_w C_w\,dt + Q_i C_i\,dt) - Q_e C\,dt + \quad 0 \quad - VCk\,dt$$

or

$$V\,dC = W\,dt - Q_e C\,dt - VCk\,dt \qquad (25\text{-}3)$$

where dC is the change in pollutant concentration in the lake, dt the incremental change in time, Q_e the outlet or effluent flow from the lake, C the average lake concentration (kg/m³), and k the first-order removal rate for the pollutant (time^{-1}). Upon rearrangement, equation (25-3) yields

$$Q_e C - W(t) + dVC = -VCk\,dt \qquad (25\text{-}4)$$

and if the Q_e, k, and V of the lake are assumed to be constant, upon rearrangement, equation (25-4) reduces to

$$V\frac{dC}{dt} + (Q_e + kV)C = W(t) \qquad (25\text{-}5)$$

If the average detention time (t_0) of the water (and thus the pollutant) in the lake is defined as

$$t_0 = \frac{V}{Q} \tag{25-6}$$

substitution and further rearrangement into equation (25-5) yields

$$V\frac{dC}{dt} + CV\left(\frac{1}{t_0} + k\right) = W \tag{25-7}$$

This is a first-order linear differential equation.

Instantaneous Pollutant Input Model

When the mass input from all sources, $W(t)$, is zero, we approach what is referred to as an *instantaneous input*. In this case, an instantaneous input is characterized as a one-time, finite addition of pollutant to the lake. For example, the release of a pollutant by a marine shipping accident would be an instantaneous input, as would a short release from an industry located on the lake. Under these conditions, integration of equation (25-7) with $W = 0$ yields

$$C(t) = C_0 e^{-[(Qe/V)+k]t} \quad \text{or} \quad C(t) = C_0 e^{-[(1/t_0)+k]t} \tag{25-8}$$

The second of equations (25-8) would be used to simulate the pollutant concentration in a lake where an instantaneous release occurred.

Step Pollutant Input Model

Next, we use equation (25-7) to derive an equation describing the constant release of a pollutant into a lake. This type of release is known as a *step input*, and an example would be the constant release from an industrial source. Under these conditions $W(t)$ is not zero (as assumed in the previous derivation), and normally there is some background concentration of pollutant in the lake system (such that C_0 in the lake cannot be considered to be zero). Here, the net pollutant concentration in the lake (and the water leaving the lake in the effluent river) is the result of two opposing forces: (1) the concentration decreases by "flushing" of the lake through the effluent river and by first-order pollutant decay, and (2) the pollutant concentration increases due to the constant input from the source. If the waste load is constant, integration of equation (25-7) yields

$$C_{(t)} = \frac{W}{\beta V}(1 - e^{-\beta t}) + C_0 e^{-\beta t} \tag{25-9}$$

where $\beta = 1/t_0 + k$ and C_0 is the background concentration of pollutant in the lake. If the background concentration in the lake is negligible, equation (25-9) reduces to

$$C_{(t)} = \frac{W}{\beta V}(1 - e^{\beta t}) \qquad (25\text{-}10)$$

These two equations can be used to estimate the concentration of pollutant in a lake that receives a constant input of pollutant. Also note that the two opposing forces described in the preceding paragraph will eventually reach equilibrium if they both remain constant. Thus, as time approaches infinity, the pollutant concentration in the lake approaches

$$C = \frac{W}{\beta V} \qquad (25\text{-}11)$$

REFERENCES

Metcalf & Eddy, Inc., *Wastewater Engineering: Collection, Treatment, Disposal*, McGraw-Hill, New York, 1972.

Serrano, S. E., *Hydrology for Engineers, Geologists, and Environmental Professionals*, Hydro-Science, Inc, Lexington, KY, 1997.

ASSIGNMENT

1. Insert the CD-ROM or install Fate on your computer (Fate is included on the CD-ROM included with your lab manual). After you have installed Fate, if it does not start automatically, open it and select the lake step or pulse module. A sample data set will load automatically. Work through the example problem, referring to the background information given earlier and the explanation of the example problem (included in Fate) as needed.

2. Select a pollutant and conduct the simulations described below for step and pulse pollution scenarios. In selecting your pollutant and input conditions, you must use a mass that will be soluble or miscible with water. An important assumption in the governing equation for all fate and transport models is that no pure solid or pure nonmiscible liquid phase of the pollutant is present.

3. Construct a pollution scenario for your simulations. This will require you to input data on a specific lake, such as the volume of the lake, inlet flow rates, outlet flow rates, background pollutant concentrations, and any pollutant decay rates (most are given in the table of first-order decay rates included in Fate).

4. Perform a simulation using your basic input data and evaluate the effluent pollutant concentration for a step and pulse pollution scenario. Next, perform a sensitivity test by selecting several input variables, such as mass loading, flow rates, or lake volume, reflecting unusually wet or dry seasons, and the first-order decay rate (those given in the table are only estimates, and the actual value can depend on factors such as volatilization, the presence of different bacterial communities, temperature, chemical degradations, photochemical degradations, etc.).

5. Finally, evaluate the assumptions of the basic model. For example, what if the entire volume of the lake was not completely mixed? How would this affect the concentration versus time plot? How would you compensate for a lake that is only 90% mixed by volume?

6. Write a three- to five-page paper discussing the results of your simulations. Include tables of data and/or printouts of figures from Fate. A copy of your report should be included in your lab manual.

To Print a Graph from Fate

For a PC

- Select the printable version of your plot (lower right portion of the screen).
- Place the cursor over the plot at the desired x and y coordinates.
- Hold the alt key down and press print screen.
- Open your print or photoshop program.

- Paste the Fate graph in your program by holding down the control key and press the letter v.
- Save or print the file as usual.

For a Mac

- Select the printable version of your plot.
- Hold down the shift and open apple key and press the number 4. This will place a cross-hair symbol on your screen. Position the cross-hair symbol in the upper right corner of your plot, click the cursor and drag the cross-hair symbol over the area to be printed or saved, release the cursor when you have selected the complete image. A file will appear on your desktop as picture 1.
- Open the file with preview or any image processing file and print it as usual.

26

FATE AND TRANSPORT OF POLLUTANTS IN GROUNDWATER SYSTEMS

Purpose: To learn two basic models for predicting the fate and transport of pollutants in groundwater systems

BACKGROUND

In this exercise we are concerned with instantaneous and step releases of a pollutant into a groundwater system. Instantaneous inputs to groundwater generally result from spills or short-term releases from pipes, tanks, or lagoons. Continuous (step) releases can occur from landfill, leaking storage tanks, and from groundwater wells. Groundwater contaminant transport, as in contaminant transport in rivers, is controlled by the physical processes of advection and dispersion. However, the causes of dispersion in a groundwater system are somewhat different from those in a river. Dispersion in groundwater systems can be broken down into microscale and macroscale processes. *Microscale variables* include molecular diffusion, pore sizes, flow path lengths, velocity gradients within flow paths, and diverging flow paths. *Macroscale dispersion* is caused by large-scale variations within the aquifer. In general, dispersion is larger in a groundwater system than in a river because of the greater number of mechanisms causing dispersion in an aquifer.

Environmental Laboratory Exercises for Instrumental Analysis and Environmental Chemistry
By Frank M. Dunnivant
ISBN 0-471-48856-9   Copyright © 2004 John Wiley & Sons, Inc.

CONCEPTUAL DEVELOPMENT OF GOVERNING FATE AND TRANSPORT EQUATION

Instantaneous Pollutant Input

Before we show the mathematical development of the governing equation for an instantaneous input, we present a conceptual approach that shows how each part of the equation relates to a physical model of an aquifer (illustrated below). First, we should note that a groundwater system is one of the most complicated environmental systems to model.

Unlike in river and lake systems modeled in Fate, pollution entering the aquifer is not mixed immediately but mixes with the groundwater as it is transported downgradient (the equivalent of downstream in a river). We handle this in the model by introducing a dispersion term, D_x. Since we are modeling only in the longitudinal (x) direction, we have only one dispersion term. If we were using a three-dimensional model, we would also need terms in the y and z directions. In addition to dispersion, most pollutants in groundwater systems react (adsorb and desorb) with the soils and minerals of the aquifer. To account for these reactions, we add a retardation term (R) calculated from the adsorption coefficient (K, described in the mathematical section below). We must also correct the volume term to account for solid particles. This is accounted for in the R term by multiplying by the bulk density (which gives an estimate of the water volume, also described in the mathematical section). We also account for chemical and biological degradation using a first-order reaction constant, k.

In the equation governing instantaneous fate and transport, we use v for the average water velocity, t for time, M for the added mass of pollutant, and x for distance from the point of introduction (usually, a groundwater well for landfill). Using this approach, we can estimate the concentration of pollutant downgradient from the point of introduction. One assumption of the model is that the pollution is added over the entire height of the porous aquifer material. In Figure 26-1, the spread of pollution downgradient is illustrated by shaded areas transitioning to larger and larger rectangles (from left to right). The increase in the size of the pollution plume is a result of mixing with the groundwater, which also dilutes the pollution and decreases the pollutant concentration. The change in shape is also a result of the adsorption/desorption phenomena and the fact that dispersion (mixing) in the x direction is the greatest. Next, we develop the model for step inputs of pollution.

Step Pollutant Input

The governing equation shown in Figure 26-2 can seem intimidating. But groundwater modeling, especially that of step inputs, is very complicated. As described in the instantaneous groundwater model, there are many chemical and physical processes that we must account for in aquifer media. The same complex dynamics of dispersion, retardation, and degradation that were discussed for

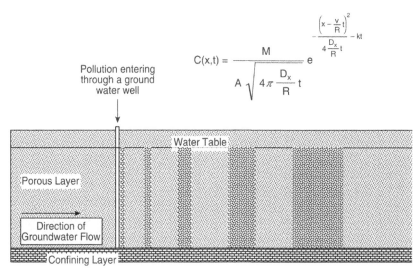

Figure 26-1. Instantaneous (pulse) input of pollution to an aquifer.

instantaneous inputs also apply to step inputs. In addition to these processes, in considering step inputs, we must account for spreading of the constantly emitted pollutant. This is completed using a mathematical error function, represented by erfc in the figure. As in the equation governing instantaneous fate and transport, we again use v for the average water velocity, t for time, C_0 for the initial concentration of pollutant, and x for distance from the point of introduction (usually, a groundwater well or landfill). Using this approach we can estimate the concentration of pollutant downgradient (as a function of distance or time) from the point of introduction. In the following figure, you will note that the pollutant plume is continuous and increases in height and diameter. You may also want to

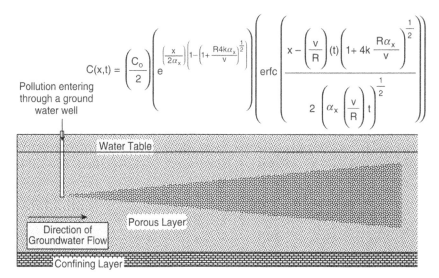

Figure 26-2. Step (continuous) input of pollution to an aquifer.

consider how the estimated pollutant concentration would change if we were using a three-dimensional model. Next, we develop the mathematical approach to groundwater modeling.

Mathematical Approach to a Lake System

Although groundwater is actually a three-dimensional system, we use a one-dimensional model in Fate to simplify the mathematics. The primary consequence of ignoring transport in the y and z directions is an underestimation of the dilution of the contaminant by spreading in these directions. The fundamental processes involved are the same in one or three dimensions.

Advection in one dimension can be described as

$$\frac{\partial C}{\partial t} = -v_x \frac{\partial C}{\partial x}$$

where C is the concentration, v_x the velocity in the x direction, t the time, and x the distance. *Dispersion* can be represented by *Fick's law* in one dimension,

$$\frac{\partial C}{\partial t} = D_x \frac{\partial^2 C}{\partial x^2}$$

where D_x is the diffusion coefficient (cm^2/s).

Chemical processes such as the biological degradation of organic compounds or the decay of radioactive compounds may also be important to the fate of groundwater contaminants. First-order degradation may be expressed as

$$\frac{dC}{dt} = -kC$$

where k is the first-order rate constant (s^{-1}) for the specific process.

If we perform a mass balance over an elemental volume of an aquifer, including the processes of advection, dispersion, and first-order chemical reaction, we obtain the equation

$$\frac{\partial C}{\partial t} = -v_x \frac{\partial C}{\partial x} + D_x \frac{\partial^2 C}{\partial x} - kC \tag{26-1}$$

Equation (26-1) is commonly referred to as the *advective–dispersive equation*. This is the same equation that governs step inputs of a contaminant to groundwater.

The most common reaction of contaminants in groundwater is *adsorption*, the attachment of a compound to a surface, is frequently modeled using a distribution coefficient, K_d:

$$K_d = \frac{S}{C}$$

where S is the concentration adsorbed (mg/g) and, C is the concentration in solution (mg/mL). The distribution coefficient assumes that the reaction is reversible and at equilibrium.

The concentration of a contaminant adsorbed to the solid phase may be described as

$$\frac{\partial S}{\partial t} = K_d \frac{\partial C}{\partial t}$$

where S is the contaminant mass on the solid phase. To convert S into mass adsorbed per elemental volume of porous media, we need to introduce bulk density, ρ_b, so that

$$\frac{\partial C^*}{\partial t} = \rho_b K_d \frac{\partial C}{\partial t}$$

where C^* is the contaminant mass on the solid phase within an elemental volume. To convert from mass per elemental volume to mass per void volume, we must incorporate porosity, n, as

$$\frac{\partial C_v}{\partial t} = \frac{\rho_b K_d}{n} \frac{\partial C}{\partial t} \tag{26-2}$$

where C_v is the of mass sorbed contaminant per void volume.

We can incorporate relationship (26-2) into the advective–dispersive equation to yield

$$\frac{\partial C}{\partial t} = -v_x \frac{\partial C}{\partial x} + D_x \frac{\partial^2 C}{\partial x^2} - \frac{\rho_b K_d}{n} \frac{\partial C}{\partial t} - kC \tag{26-3}$$

Equation (26-2) can be rearranged to yield

$$\frac{\partial C}{\partial t} \left(1 + \frac{\rho_d K_d}{n} \right) = -v_x \frac{\partial C}{\partial x} + D_x \frac{\partial^2 C}{\partial x^2} - kC$$

or

$$R \frac{\partial C}{\partial t} = -v_x \frac{\partial C}{\partial x} + D_x \frac{\partial^2 C}{\partial x^2} - kC \tag{26-4}$$

The term $1 + \rho_b K_d/n$ is called the retardation factor, R. The retardation factor represents the retardation of the solute relative to the average groundwater velocity (v), or

$$R = \frac{v}{v_c}$$

where v_c is the contaminant velocity and v is the groundwater velocity. When $v = v_c$, $R = 1$ and the contaminant is said to be conservative (i.e., it does not adsorb to the solid and has a K_d value of 0).

Instantaneous Pollutant Input

If we assume that the spill contaminates the entire thickness of the aquifer, equation (26-4) can be integrated to yield

$$C(x,t) = \frac{M}{A\sqrt{4\pi(D_x/R)t}} \exp\left\{ -\frac{[x - (v/R)t]^2}{A(D_x/R)t} - kt \right\}$$

where $x = $ distance from the source
   $t = $ time
   $M = $ mass of contaminant added to the aquifer
   $A = $ cross-sectional void volume contaminated by the pollution
   $D_x = $ dispersion coefficient
   $R = $ retardation factor
   $v = $ velocity
   $k = $ first-order reaction rate

Step Pollutant Input

For the initial condition $C(x,0) = 0$, where the concentration equals zero everywhere, and the boundary condition $C(0,t) = C_0$, where the concentration at the source remains constant at the value of C_0, the advective–dispersive equation may be solved using Laplace transformations to yield

$$C(x,t) = \frac{C_0}{2} \exp\left\{ \frac{x}{2\alpha_x}\left[1 - \left(1 + \frac{R \cdot 4k\alpha_x}{v}\right)^{1/2}\right] \right\}$$
$$\cdot \left\{ \mathrm{erfc}\left[\frac{x - (v/R)(t)[1 + 4k(R\alpha_x/v)]^{1/2}}{2[\alpha_x(v/R)t]^{1/2}}\right] \right.$$
$$\left. + e^{x/\alpha_x}\,\mathrm{erfc}\left[\frac{x + (v/R)(t)[1 + 4k(R\alpha_x/v)]^{1/2}}{2[\alpha_x(v/R)t]^{1/2}}\right] \right\}$$

where $C_0 = $ initial concentration of the contaminant
   $x = $ distance from the source
   $\alpha_x = $ longitudinal dispersivity
   $k = $ first-order reaction rate
   $v = $ velocity
   $t = $ time
   erfc $= $ complementary error function

The final term in equation (26-5),

$$e^{x/\alpha_x} \; \text{erfc} \left\{ \frac{x + (v/R)t[1 + 4k(R\alpha_x/v)]^{1/2}}{2[\alpha_x(v/R)t]^{1/2}} \right\}$$

is generally considered insignificant and is ignored; the term is also ignored in Fate.

Finally, we discuss two terms in the final fate and transport equations. Dispersion in groundwater, as in rivers, is a function of velocity, or

$$D = \alpha_x v$$

where α_x is the called the *dispersivity*. Because dispersivity is a function only of the aquifer matrix and not of velocity, it is used in many groundwater models in preference to the dispersion coefficient. Because of the many causes of dispersion discussed previously, dispersivity is one of the most difficult parameters to measure accurately. Dispersivity values tend to increase with the scale over which they were measured because the degree of heterogeneity within the aquifer generally increases with the scale.

The *error function* is the area between the midpoint of the normal curve and the value for which you are taking the error function. The *complementary error function*, the error function subtracted from 1, accounts for the spreading of the plume.

REFERENCES

Fetter C. W., *Applied Hydrogeology*, Charles E. Merrill, Toronto, 1980.

Fetter C. W., *Contaminant Hydrogeology*, Macmillan, New York, 1993.

ASSIGNMENT

1. Install Fate on your computer (Fate is included with your lab manual). Open the program and select the groundwater step or pulse module. A sample data set will load automatically. Work through the example problem, referring to the background information above and the explanation of the example problem (included in Fate) as needed.

2. Select a pollutant and conduct the simulations described below for step and pulse pollution scenarios. In selecting your pollutant and input conditions, you must use a mass that will be soluble or miscible with water. An important assumption in the governing equation for all fate and transport models is that no pure solid or pure nonmiscible liquid phase of the pollutant is present.

3. Construct a pollution scenario for your simulations. This will require you to insert data on a specific aquifer, such as the volume of the system, groundwater flow rates, background pollutant concentrations (usually assumed to be zero), adsorption coefficients (K), dispersivity values, and any pollutant decay rates (most are given in the table of first-order decay rates included in Fate).

4. Perform a simulation using your basic input data, and evaluate the downgradient pollutant concentration for the step and pulse pollution scenarios (as a function of time and distance). Next, perform a sensitivity test by selecting and varying input variables, such as mass loading, flow rate or bulk density, K values, and first-order decay rate (those given in the table are only estimates, and the actual value can depend on factors such as the present of different bacterial communities, temperature, chemical degradations, etc.).

5. Finally, evaluate the assumptions of the basic model. For example, what if you use a three-dimensional model? How will your downgradient concentration values differ?

6. Write a three- to five-page paper discussing the results of your simulations. Include tables of data and/or printouts of figures from Fate. A copy of your report should be included in your lab manual.

To Print a Graph from Fate

For a PC

- Select the printable version of your plot (lower right portion of the screen).
- Place the cursor over the plot at the desired x and y coordinates.
- Hold the alt key down and press print screen.
- Open your print or photoshop program.
- Paste the Fate graph in your program by holding down the control key and press the letter v.
- Save or print the file as usual.

For a Mac

- Select the printable version of your plot.
- Hold down the shift and open apple key and press the number 4. This will place a cross-hair symbol on your screen. Position the cross-hair symbol in the upper right corner of your plot, click the cursor and drag the cross-hair symbol over the area to be printed or saved, release the cursor when you have selected the complete image. A file will appear on your desktop as picture 1.
- Open the file with preview or any image processing file and print it as usual.

27

TRANSPORT OF POLLUTANTS IN THE ATMOSPHERE

Purpose: To learn two basic models for predicting the fate and transport of pollutants in atmospheric systems

BACKGROUND

The atmosphere is the environmental medium where we live and breath. Modeling of atmospheric pollution can be used to determine human exposure to existing pollution sources and to predict future exposures from industrial accidents. There are many sources of atmospheric pollution, including volcanoes, industrial smoke stacks, fugitive (or nonpoint) industrial emissions, gasoline stations, forest fires, industrial accidents, and automotive and railroad accidents. In Fate, we develop relatively simple models to predict the fate and transport of pollution released such sources.

First, we compare other fate and transport models to the general atmospheric model. The aquatic models in Fate were given only for one or two dimensions. Streams and lakes can be modeled adequately using one-dimensional models since most of the dispersion is in the longitudinal direction, whereas groundwater systems require at least two dimensions (x and y). Two dimensions are required in the latter system because the groundwater is not constrained by a river or lake bank, and dispersion can occur in all directions. Vertical dispersion, although important near a point pollution source, becomes less important when the

Environmental Laboratory Exercises for Instrumental Analysis and Environmental Chemistry
By Frank M. Dunnivant
ISBN 0-471-48856-9 Copyright © 2004 John Wiley & Sons, Inc.

groundwater system is bounded by confining layers above and below the aquifer of interest, which is why we used the simpler two-dimensional model in the instantaneous and pulse groundwater releases.

Although the aquatic models may have seemed complicated, they are simpler than most atmospheric models. Because of wind currents and mixing, atmospheric models have to incorporate three dimensions, which automatically makes the governing equations more complex. As usual, we make many assumptions that make our model more manageable. For example, the models given in Fate are not designed for gases that are more or less dense than the atmosphere, and therefore ignore buoyancy effects. The models distinguish between step and instantaneous sources, although actual atmospheric pollution episodes can lie between these two extremes. Unlike the aquatic models that allow first-order decay processes, our atmospheric models do not allow degradation of pollutants. This assumption is justified for models of a pollutant over relatively short distances (under 10,000 meters or 7 miles) because most photochemical reactions (except for the production of smog) require the pollutant to be in the atmosphere over a much longer time frame (hours to days). The dominant force resulting in the reduction of the pollutant concentration is dispersion, which can dilute pollutant concentrations rapidly. However, understanding and accounting for dispersion can be very complicated. First, we look at the movement of atmospheric gases over Earth's surface.

A profile of the wind's velocity with increasing height has a steep increasing parabolic shape, with low velocity at Earth's surface due to friction between the moving air and the ground. The surface wind velocity is also subject to many complex variables, however. For example, the roughness of Earth's surface can significantly affect the shape or steepness of the wind velocity–height profile. The wind velocity profile over an open grassland is illustrated on the right-hand side of Figure 27-1, showing that wind speed approaches its maximum rapidly as height

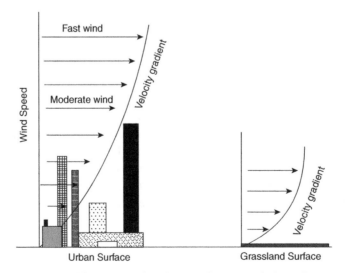

Figure 27-1. Effect of surface roughness on wind speed.

above the surface increases. Compare this to an urban setting, where tall buildings impede the path of the wind and slow its speed. This expands the velocity–height gradient well above Earth's surface. The resulting lower wind velocity could decrease the turbulence and subsequent dispersion by slowing the wind velocity but may also result in stagnant pockets of the atmosphere that can contain clear or polluted air. Thus, the increase in the surface's roughness from the presence of buildings will greatly affect flow patterns and ground-level pollutant concentrations. Variables such as this demonstrate that atmospheric processes are too complicated even for our most sophisticated models. In our brief introduction we simplify our model by assuming that an average wind speed can be used and, in general, we do not account for differences in surface roughness.

Although surface roughness can greatly affect turbulence and mixing, the magnitude of wind speed can also increase mixing. We refer to this mixing as *dispersion*, since the net result is a dilution of pollutant concentrations. If we combine the effects of wind velocity and atmospheric temperature as a function of height above the surface, we obtain the three basic turbulence scenarios shown in Figure 27-2. We start with an isolated pocket of atmosphere at nighttime temperatures (shown in Figure 27-2a). This type of condition occurs where a thick cloud layer prevents Earth from radiating its heat to space as it cools during the night. Under these conditions, an emission from an industrial stack will take the shape of the plume shown in Figure 27-2a. The gases released will rise or sink until their density (temperature) matches that of the surrounding (diluting) atmospheric gases. Then the plume will take the shape of a thin layer.

Under daytime heating conditions, the temperature–height profile will be similar to that shown in Figure 27-2b. In a steady wind, the plume will spread in all directions, but primarily in the longitudinal direction. With a lower

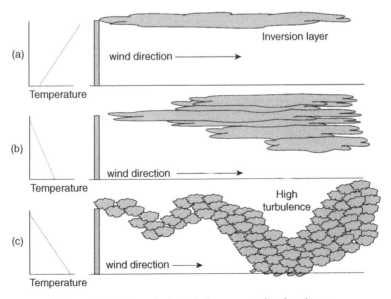

Figure 27-2. Three basic turbulence scenarios for plumes.

TABLE 27-1. Pasquill Stability Categories

Windspeed at 10 m Elevation (m/s)	Day, Degree of Cloud Insolation			Night	
				Thinly Overcast or Greater Than 50% Low clouds	Less Than 50% Cloud Cover
	Strong	Moderate	Slight		
< 2	A	A, B	B	G	G
2–3	A, B	B	C	E	F
3–5	B	B, C	D	D	E
5–6	C	C, D	D	D	D
>6	C	D	D	D	D

Source: Turner (1994) and Pasquill (1961). Turner (1994) adds the following notes on selecting the appropriate category:

1. Strong insolation corresponds to sunny midday in midsummer in England; slight isolation to similar conditions in midwinter.
2. Night refers to the period from 1 hour before sunset to 1 hour after sunrise.
3. The neutral category D should also be used, regardless of wind speed, for overcast conditions during day or night and for any sky condition during the hour preceding or following night as defined in note 2.

temperature–height gradient and a higher wind velocity, extreme turbulence will be observed (Figure 27-2c). To attempt the modeling of these conditions, we must greatly simplify the temperature and wind relationships.

We start our simplification process by attempting to combine the effects of wind velocity, temperature–height profiles, and cloud cover into a set of atmospheric stability categories. As we do this, remember that our goal is to come up with a way to characterize dispersion (mixing) of the pollutant with the atmospheric gases. Table 27-1 shows a qualitative approach to the combined effects of wind speed and cloud cover collected for rural settings in England. Cloud cover is a good reflection of heat back to Earth. The categories range from strongly unstable (category A, reflected in Figure 27-2c) to very stable (category G) and distinguish between day and night conditions.

Next, the somewhat qualitative categories in Table 27-1 are used to predict values for horizontal dispersion coefficients (Table 27-2), which are estimates of mixing in the x and y directions. We do not have a way mathematically to predict these values accurately, and the data in Tables 27-1 and 27-2 are empirical (based on experimental observations). We usually assume that dispersion in the x and y directions is the same; thus Table 27-2 can be used to estimate σ_x and σ_y simultaneously. The equations given in Table 27-1 were used to draw the lines in Figure 27-3. Note that dispersion increases as you move away from the point source of pollution. This should be intuitive, since mixing continues and the wind causes more mixing as you move away from the point source. So for every pollutant concentration you attempt to estimate, you must select a distance from the point source. The unfortunate result of this is that Fate can only plot a slice of

**TABLE 27-2. Pasquill–Gifford Horizontal
Dispersion Parameters**

$$s_y = 1000 \times \tan T/2.15$$

where x is the downwind distance (in kilometers) from the
point source and T, which is one-half Pasquill's q in degrees
T as a function of x, is determined by each stability category
in Table 27-1.

Stability	Equation for T
A	$T = 24.167 - 2.5334 \ln x$
B	$T = 18.333 - 1.8096 \ln x$
C	$T = 12.5 - 1.0857 \ln x$
D	$T = 8.333 - 0.7238 \ln x$
E	$T = 6.25 - 0.5429 \ln x$
F	$T = 4.167 - 0.3619 \ln x$

Source: Turner (1994).

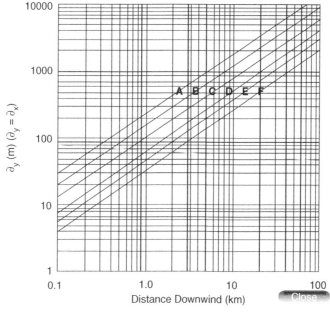

Figure 27-3. Pasquill–Gifford horizontal dispersion parameters. (From Turner, 1970; Pasquill,
1961.)

the concentration in the y and z planes. You will have to plot manually the
concentration gradient in the x, or longitudinal, direction.

Dispersion in the vertical (z) direction is somewhat more complicated to
predict and again is based on experimental observations. We can estimate the
vertical dispersion coefficient, σ_z, by using the same atmospheric stability
categories from Table 27-1 but with a more precise treatment of the wind

speed. The equation governing the estimate of vertical dispersion is

$$\sigma_z = ax^b$$

where x is the distance in kilometers and a and b are fitting parameters obtained from Table 27-3.

TABLE 27-3. Pasquill–Gifford Vertical Dispersion Parameter[a]

Stability	Distance (km)	a	b	s_z at Upper Boundary
A	>3.11			5000
	0.5–3.11	453.85	2.1166	
	0.4–0.5	346.75	1.7283	104.7
	0.3–0.4	258.89	1.4094	71.2
	0.25–0.3	217.41	1.2644	47.4
	0.2–0.25	179.52	1.1262	37.7
	0.15–0.2	170.22	1.0932	29.3
	0.1–0.15	158.08	1.0542	21.4
	<0.1	122.8	0.9447	14.0
B	>0.35			5000
	0.4–35	109.30	1.0971	
	0.2–0.4	98.483	0.9833	40.0
	>0.2	90.673	0.93198	20.2
C	all values of x		61.141	0.91465
D	>30	44.053	0.51179	
	10–30	36.650	0.56589	251.2
	3–10	33.504	0.60486	134.9
	1–3	32.093	0.64403	65.1
	0.3–1	32.093	0.81066	32.1
	<0.3	34.459	0.86974	12.1
E	>40	47.618	0.29592	
	20–40	35.420	0.37615	141.9
	10–20	26.970	0.46713	109.3
	4–10.	24.703	0.50527	79.1
	2–4	22.534	0.57154	49.8
	1–2	21.628	0.63077	33.5
	0.3–1	21.628	0.75660	21.6
	0.1–0.3	23.331	0.81956	8.7
	<0.1	24.260	0.83660	3.5
F	>60	34.219	0.21716	
	30–60	27.074	0.27436	83.3
	15–30	22.651	0.32681	68.8
	7–15	17.836	0.4150	54.9
	3–7	16.187	0.4649	40.0
	2–3	14.823	0.54503	27.0
	1–2	13.953	0.63227	21.6
	0.7–1	13.953	0.68465	14.0
	0.2–0.7	14.457	0.78407	10.9
	<0.2	15.209	0.81558	4.1

Source: Turner (1970); Pasquill (1961).
[a] $\sigma_z = ax^b$, where x is in kilometers.

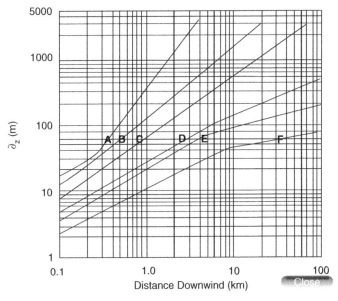

Figure 27-4. Pasquill–Gifford vertical dispersion parameters. (From Turner, 1970.)

A plot of the dependence of vertical dispersion coefficients on distance from the point source is shown in Figure 27-4. We have been describing dispersion, but what exactly is it? As we have noted, dispersion is a function of the distance from the point source. Dispersion is a mathematical description of mixing between the pollutant plume and the natural atmospheric gases. The values you read from the graph or calculate using the equations are given in meters or kilometers. Thus, the values given represent the width of the pollutant plume at the specified distance from the point source and thus reflect the amount of atmosphere with which the pollution has mixed.

STEP INPUT (PLUME MODEL) OF POLLUTANT

Using the many assumptions stated earlier and the estimated horizontal and vertical dispersion coefficients, the *plume model* [equation (27-1)] can be derived, using differential equation techniques to estimate the pollutant concentration at any point (x, y, and z) downwind from the continuous source:

$$C(x,y,z) = \frac{Q_m}{2\pi\sigma_y\sigma_z u} \left[\exp -\frac{1}{2}\left(\frac{y}{\sigma_y}\right)^2 \right] \left\{ \exp\left[-\frac{1}{2}\left(\frac{z-H_r}{\sigma_z}\right)^2 \right] + \exp\left[-\frac{1}{2}\left(\frac{z+H_r}{\sigma_z}\right)^2 \right] \right\}$$

$$(27\text{-}1)$$

where $C(x, y, z)$ = concentration of pollutant in the plume as a function of x, y, and z (mass/length^3)

x, y, z = are distances from the source (length) (see Figures 27-3 and 27-4)

Q_m = pollutant source (mass/time)

$\sigma_x = \sigma_y$ = horizontal dispersion coefficient (length)

σ_z = vertical dispersion coefficient (length)

u = wind velocity (length/time)

H_r = height of the release (length)

Notice the terms that we need to use this mode: the mass of pollutant released, the wind speed, the x, y, and z coordinates that yield estimates of dispersion (mixing), and the height of the release above Earth's surface. All of these are relatively simple to estimate using the techniques described earlier.

For the concentration along the centerline of the plume ($z = 0$ and $H_r = 0$), we can use a simplification of equation (27-1):

$$C(x, y, 0) = \frac{Q_m}{\pi \sigma_y \sigma_z u} \left[\exp - \frac{1}{2} \left(\frac{y}{\sigma_y} \right)^2 \right] \tag{27-2}$$

A typical simulation of downwind pollutant concentration is shown in Figure 27-5 for a 1.0-m z value (height above ground level), a y distance of 0.0 km (along the x–z axis), and an x value (distance downwind) of 1.5 km. In Figure 27-5 the height of the Gaussian-shaped plot is along the center x axis (a y value of zero) and 1 m

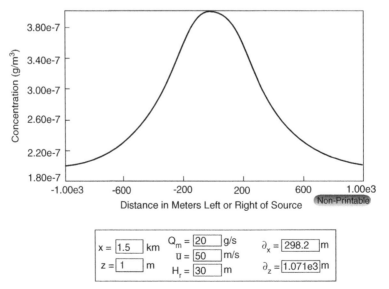

Figure 27-5. Output from Fate for a continuous release (plume) of pollutant into the atmosphere as you look along the x-axis.

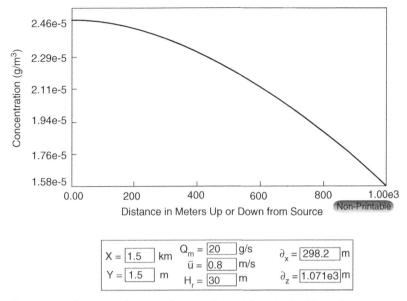

Figure 27-6. Output from Fate for a continuous release (plume) of pollutant into the atmosphere showing variations in plume concentration with changing vertical position in relation to the source.

above ground or about nose level for a tall person. The pollutant concentration declines as you go to the left or right of the centerline (an increase or decrease of y values). Note that the width of the main plume concentration covers a range of approximately 1200 m (from -600 m to the left to $+600$ m to the right).

A similar output would be obtained by plotting a y value of 0.0 (along the centerline), an x distance of 1.5 km, and calculating the pollutant concentration as you move up in the atmosphere. This is illustrated in Figure 27-6. In this plot, as you go from left to right on the x axis, you are moving up in the atmosphere.

Another useful function of Fate is to evaluate the pollutant concentration as a function of distance from the point source. Fate cannot plot this directly since dispersion in the x, y, and z directions are a function of distance from the point source. To accomplish this we must repeatedly use steps 5 and 6 in the plume model. Change the x distance systematically, increase it incrementally, and record the pollutant concentration given in step 6. A plot like the one shown in Figure 27-7 can be obtained. Note that the pollutant concentration decreases, as expected, as you move away from the point source.

PULSE INPUT (PUFF MODEL) OF POLLUTION

For a pulse rather than a plume input, dispersion is handled a little differently. In the step (plume) model we can use either rural or urban dispersion estimates, whereas urban dispersion parameters are generally used for the pulse (puff) model. These dispersion estimates are derived from experimental observations

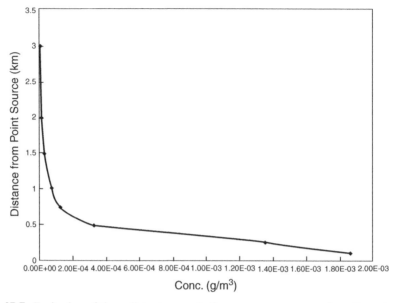

Figure 27-7. Evaluation of the pollutant concentration as you move away from the point source (plume model).

made by McElroy and Pooler (1968) near St. Louis and from Briggs (1972). Calculations for estimating the horizontal and vertical dispersion coefficients are given in Table 27-4. We again assume that dispersion in the x and y directions are the same. Atmospheric stability categories are the same as those described in Table 27-1. Vertical and horizontal dispersion coefficients are shown in Figures 27-8 and 27-9, respectively.

TABLE 27-4. Dispersion Parameters[a]

Pasquill Type of Stability	s_y (m)	s_z (m)
Urban		
A, B	$0.32/(0.0004x)^{-0.5}$	$0.24/(0.001x)^{0.5}$
C	$0.22/(0.0004x)^{-0.5}$	$0.20x$
D	$0.16/(0.0004x)^{-0.5}$	$0.14/(0.0003x)^{-0.5}$
E, F	$0.11/(0.0004x)^{-0.5}$	$0.08/(0.0015x)^{-0.5}$
Open-Country[b]		
A	$0.22x/(1 + 0.0001x)^{0.5}$	$0.20x$
B	$0.16x/(0.0001x)^{0.5}$	$0.12x$
C	$0.11x/(1 + 0.0001x)^{0.5}$	$0.08x/(1 + 0.0002x)^{0.5}$
D	$0.08x/(1 + 0.0001x)^{0.5}$	$0.06x/(1 + 0.0015x)^{0.5}$
E	$0.06x/(1 + 0.0001x)^{0.5}$	$0.03x/(1 + 0.0003x)$
F	$0.04x\ (1 + 0.0001x)^{0.5}$	$0.016x/(1 + 0.0003x)$

Source: Turner (1994); Briggs (1972); McElroy and Pooler (1968).

[a] For distances x between 100 and 10,000 m.

[b] Not used in Fate, but you may enter the calculated values manually.

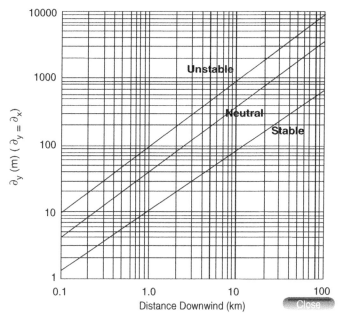

Figure 27-8. Pasquill–Gifford vertical dispersion parameters. (From Turner, 1970.)

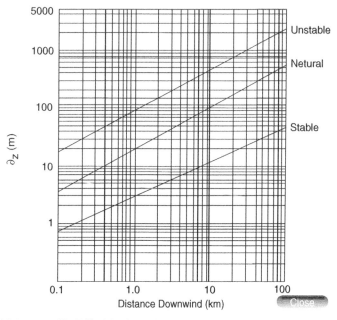

Figure 27-9. Pasquill–Gifford horizontal dispersion parameters. (From Turner, 1970.)

Using the stability categories, wind speed, and the equations shown in Table 27-4, we can now estimate the atmospheric pollutant concentration downwind from an instantaneous (also referred to as *pulse* or *puff*) source by

$$C(x,y,z,t) = \frac{Q_m}{(2\pi)^{3/2}\sigma_x\sigma_y\sigma_z} \exp\left[-\frac{1}{2}\left(\frac{y}{\sigma_y}\right)^2\right]\left\{\exp\left[-\frac{1}{2}\left(\frac{z-H_r}{\sigma_z}\right)^2\right]\right.$$

$$\left. + \exp\left[-\frac{1}{2}\left(\frac{z+H_r}{\sigma_z}\right)^2\right]\right\}$$

(27-3)

where $C(x,y,z,t)$ = concentration of pollutant in the plume as a function of x, y, and z (mass/length^3)

x, y, z = distances from the source (length) (see Figures 27-3 and 27-4)

t = time

Q_m = pollutant source (mass/time)

σ_x, σ_y = horizontal dispersion coefficients (length)

σ_z = vertical dispersion coefficient (length)

H_r = height of the release (length)

Note the inclusion of time since the distance traveled (x) is a function of wind velocity (u) and time (t), where

$$x = ut$$

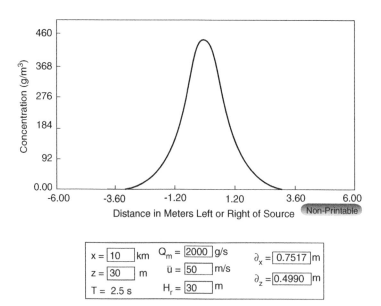

Figure 27-10. Output from Fate for a pulse release (puff) of pollutant into the atmosphere with variation in horizontal distance from source.

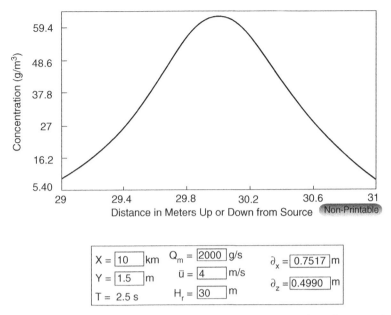

Figure 27-11. Output from Fate for a pulse release (puff) of pollutant into the atmosphere with variation in vertical distance from source.

For the concentration along the centerline ($y = 0$, $z = 0$, and $H_r = 0$) we can use a simplification of equation (27-3), to yield

$$C(x, 0, 0, t) = C(ut, 0, 0, t) = \frac{Q_m}{(\sqrt{2}\,\pi)^{3/2}\sigma_x\sigma_y\sigma_z} \qquad (27\text{-}4)$$

Simulation output from Fate is shown in Figures 27-10 and 27-11 for viewing pollutant concentration along the y and the z axes, respectively.

REFERENCES

Briggs, G. A., *Atmos. Environ.*, **6**, 507–510 (1972).

Gifford, F. A., *Nucl. Safety*, **17**(1), 68–86 (1976).

McElroy, J. L. and F. Pooler, St. Louis dispersion study, U.S. Public Health Service, National Air Pollution Control Administration Report AP-53, 1968.

Pasquill, F., *Meterol. Mag.*, **90**(1063), 33–49 (1961).

Pasquill, F., *Atmospheric Dispersion Parameters in Gaussian Plume Modeling: Part II. Possible Requirements for Change in the Turner Workbook Values*. EPA-600/4-76-030b. U.S. Environmental Protection Agency, Research Triangle Park, NC, 1976.

Turner, D. B., *Workbook of Atmospheric Dispersion Estimates*, Department of Health, Education, and Welfare, Cincinnati, OH, 1970.

Turner, D. B., *Workbook of Atmospheric Dispersion Estimates: An Introduction to Dispersion Modeling*, 2nd ed., Lewis Publishers, Ann Arbor, MI, 1994.

ASSIGNMENT

1. Install Fate on your computer (Fate is included on the CD-ROM with your lab manual). After you have installed Fate, if it does not start automatically, open it and select the air step or pulse module. A sample data set will load automatically.

2. Select a pollutant and conduct the simulations described below for both step and pulse pollution scenarios. Construct a pollution scenario for your simulations. This will require you to provide data on specific atmospheric conditions such as pollutant release rates and wind speed.

3. Perform a simulation using your basic input data and evaluate the down-gradient pollutant concentration for a step and pulse pollution scenario. Next, perform a sensitivity test to study the effect of wind velocity on downgradient pollutant concentrations. You will have to do this manually using steps 5 and 6 in Fate and use a spreadsheet to compile your results.

4. Write a three- to five-page paper discussing the results of your simulations. Include tables of data and/or printouts of figures from Fate. A copy of your report should be included in your lab manual.

To Print a Graph from Fate

For a PC

- Select the printable version of your plot (lower right portion of the screen).
- Place the cursor over the plot at the desired x and y coordinates.
- Hold the alt key down and press print screen.
- Open your print or photoshop program.
- Paste the Fate graph in your program by holding down the control key and press the letter v.
- Save or print the file as usual.

For a Mac

- Select the printable version of your plot.
- Hold down the shift and open apple key and press the number 4. This will place a cross-hair symbol on your screen. Position the cross-hair symbol in the upper right corner of your plot, click the cursor and drag the cross-hair symbol over the area to be printed or saved, release the cursor when you have selected the complete image. A file will appear on your desktop as picture 1.
- Open the file with preview or any image processing file and print it as usual.

28

BIOCHEMICAL OXYGEN DEMAND AND THE DISSOLVED OXYGEN SAG CURVE IN A STREAM: STREETER–PHELPS EQUATION

Purpose: To learn a basic model (the Streeter–Phelps equation) for predicting the dissolved oxygen concentration downstream from an organic pollution source

BACKGROUND

One of the greatest environmental accomplishments is sanitary treatment of most human waste (sewage). Improper treatment of these wastes has led to outbreaks of cholera, typhoid, and other human-waste-related diseases and many human deaths worldwide (see Chapter 19). Today, most developed nations have greatly minimized or eliminated the spread of these diseases through treatment of sewage waste. In general, our efforts to minimize the effects of these wastes can be divided into two approaches. First, sewage is treated in engineered systems such as sewage treatment plants, where large amounts of waste enter the system and are treated prior to release. However, it is only economical to treat or remove approximately 95 to 98% of the original organic matter entering the treatment plant. After removal of pathogenic organisms, the remaining organic matter is then released to an adjacent natural water body, where the remaining organic

Environmental Laboratory Exercises for Instrumental Analysis and Environmental Chemistry
By Frank M. Dunnivant
ISBN 0-471-48856-9   Copyright © 2004 John Wiley & Sons, Inc.

matter is oxidized slowly as it is transported down the system. When the treatment plant is designed properly and under normal conditions, natural systems can handle these small amounts of waste and undergo self-purification. Self-purification is a process that nature uses every day to recycle nutrients in watersheds, specifically carbon and nitrogen.

Because the degradation of organic matter consumes oxygen that is dissolved in the stream water, we describe organic waste in terms of how much oxygen is needed to degrade (or oxidize) the waste. This is referred to as the *biochemical oxygen demand* (BOD). When waste enters a system faster than it can be degraded, dissolved oxygen levels can drop below the minimum level required by aquatic organisms. In extreme cases, all of the dissolved oxygen may be removed, making the stream "*anoxic*". When this happens, most organisms die, thus adding more BOD to the system and further increasing the oxygen demand.

Organic matter in the form of human waste, animal waste, or decaying components of nature exerts BOD on natural systems. Lakes and streams can be characterized in terms of the amount of organic matter in the system. If too much organic matter is present, the system may go anoxic during certain periods of the day or year. For example, streams can experience diurnal cycles with high dissolved oxygen (O_2) concentrations during the day when photosynthesis is occurring, and low O_2 concentrations during the night when respiration and decay processes dominate. Lakes usually experience annual cycles, with anoxic conditions occurring in the bottom of lakes during the summer months. The goal in wastewater engineering is to remove sufficient amounts of the BOD (it is virtually impossible to remove all of the BOD) such that the natural receiving body of water (i.e., stream or lake) can self-purify the system and avoid developing anoxic regions in the system. Modern sewage treatment facilities generally remove greater more than 95% of the oxidizable organic matter. However, there are many aging facilities in the United States that do not meet these requirements. In addition, facilities in metropolitan areas have combined storm and sanitary systems and during periods of flooding routinely exceed the capacity of the sewage treatment plant. When this happens, a portion (or all) of the combined waste from the sewer system bypasses the sewage treatment and enters the receiving body of water untreated. This allows anoxic zones to develop in the natural system and possibly increases the transmission of disease-causing agents. Another major type of BOD release to the natural system comes from stock farming operations where grazing pastures, feedlots, or stockyards are allowed to drain directly into a receiving water body. Each of the situations described above can lead to oxygen depletion in natural water bodies. The resulting oxygen level, as a function of distance from the source, can be estimated using the equations derived below. The goal of these calculations is to provide the user with an estimate of the shape of the dissolved oxygen curve, the minimum oxygen concentration and the distance from the source where the lowest dissolved oxygen concentration will occur, and the concentration of dissolved oxygen at any distance from the source.

CONCEPTUAL DEVELOPMENT OF THE GOVERNING FATE AND TRANSPORT EQUATION

There are several assumptions that we must make to develop a relatively simple equation for calculating the dissolved oxygen in a stream containing organic waste [equation (28-2)]. For example, we assume that the waste is applied evenly across the width of the stream and that it is instantly mixed with the stream water. Of course, we need to know the waste and stream flow rates and the concentration of BOD in the waste (BOD_L in the governing equation). The two necessary kinetic parameters are the rate at which oxygen is consumed by microorganisms (k_2') and the rate at which oxygen is readded to the stream from the atmosphere (k'). Each of these kinetic terms is dependent on diffusion and is therefore exponential in nature (represented by the e term in the governing equation). The final quantity we need is the dissolved oxygen content of the stream above the point of waste entry (D_0). The additional terms x and v in equation (28-2) represent the distance downstream from the waste inlet and the velocity of the stream water, respectively.

$$D = \frac{k' \cdot BOD_L}{k_2' - k'} \left(e^{-k'(x/v)} - e^{-k_2'(x/v)} \right) + D_0 e^{-k_2'(x/v)}$$

Notice the shape of the dissolved oxygen curve in Figure 28-1. Above the inlet of wastewater the dissolved oxygen (DO at $x = 0$) is high and near the water saturation value. As organic waste enters the stream, the DO declines sharply, initially due to the mixing of clean oxygenated water with sewage effluent and later due to the consumption of oxygen by microorganisms. The curve reaches a minimum DO concentration, referred to as the *critical point*, and slowly increases

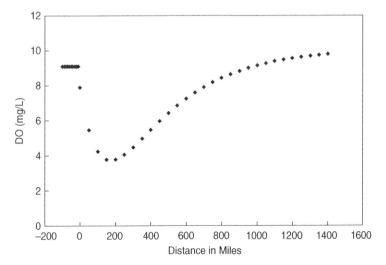

Figure 28-1. Typical dissolved oxygen sag curve for a polluted stream.

back to the original DO concentration seen above the input of waste to the stream. Next we look more closely at the mathematical derivation of the governing equation.

MATHEMATICAL APPROACH TO A LAKE SYSTEM

The governing equation used to estimate the dissolved oxygen concentration in stream water is derived by taking a mass balance of BOD in the system, such that

change in BOD concentration	=	inflow of BOD to the stream segment	−	outflow of BOD from the stream segment	+	other sources of BOD	−	losses of BOD

Flow through a cross section of the stream channel can be described mathematically as

$$V \Delta C = QC \Delta t - Q\left(C + \frac{\partial C}{\partial x}\Delta x\right)\Delta t + 0 - VkC \Delta t \qquad (28\text{-}1)$$

where V is the volume of water in the cross section containing the waste, ΔC the change in BOD concentration, Q the flow rate of water containing BOD into and out of the cross section of the channel, Δt the change in time, C the average concentration of BOD in the cross section, and $\partial C/\partial x$ the rate of change of BOD concentration with change in distance from the point source.

Note that each term in these equations are in units of mass, hence the name *mass balance*. If each side of the equation is divided by Δt, we obtain

$$V \frac{\Delta C}{\Delta t} = -Q\frac{\partial C}{\partial x}\partial x - kVC$$

Metcalf & Eddy (1972) show how the concentration (C) of BOD can be expressed in terms of mg O_2/L and integrate the new equation to obtain a relatively simple equation that can be used to predict oxygen concentration any distance downstream from the source for a relatively rapidly moving stream (one basic assumption is that there will be no settling of sewage along the bottom of the stream channel). This equation can be represented by

$$D = \frac{k' \cdot \mathrm{BOD}_L}{k_2' - k'}\left(e^{-k'(x/v)} - e^{-k_2'(x/v)}\right) + D_0\, e^{-k_2'(x/v)} \qquad (28\text{-}2)$$

where D = dissolved oxygen concentration (mg O_2/L)
$\quad k'$ = BOD rate constant for oxidation (day^{-1})
BOD_L = ultimate BOD (mg/L)
$\quad k_2'$ = reaeration constant (to the base e, day^{-1})

x = distance from the point source (miles or kilometers)

v = average water velocity (miles/day or kilometers/day, but units must be compatible with distances, x)

D_0 = initial oxygen deficit (mg/L)

Note the introduction of a few new terms. The term k' is the first-order rate constant associated with reaeration of the stream water. Exact measurement of this parameter is difficult since it is dependent on factors such as stream depth, mixing in the stream, and degree of water and air contact. For simplification purposes, a set of values has been tabulated by the Engineering Board of Review for the Sanitary District of Chicago (1925) and can be used based on a qualitative description of the stream. These values have been summarized by Metcalf & Eddy (1972) and are given in Table 28-1. Note that for k values, the log to the base e (natural log) must be used in all calculations.

The second term, BOD_L, is the ultimate BOD or maximum oxygen required to oxidize the waste sample completely. This value is also determined or estimated through the BOD experiment. Normally, BOD values are determined on a five-day basis, which corresponds to the O_2 consumed during the first five days of degradation. However, since we may be concerned with a travel time in the stream exceeding five days, we need to know the ultimate BOD (BOD_L). This value can be determined experimentally or estimated from the BOD_5 value using the equation

$$BOD_L = \frac{BOD_5}{1 - e^{-k'(x/v)}} \qquad (28\text{-}3)$$

The k'_2 term is the reaeration constant and is specific to the stream of interest. This is obtained by conducting an oxygen uptake experiment known as a *BOD experiment*, in which a set of diluted wastewater samples are saturated with oxygen, sealed, and sampled to determine how much oxygen remains as a function of time. The plot of the data (oxygen consumed, in milligrams, versus time, in days) is exponential, and the curvature of the plot can be described by the rate constant, k', in day^{-1}.

For examples and calculations, the distance downstream from the BOD source, x, can be given in miles or kilometers, but units must be consistent. It should be

TABLE 28-1. Reaeration Constants

Water Body	Ranges of k'_2 at 20°C (Base 10)	Ranges of k'_2 at 20°C (Base e for Calculations)
Small ponds and backwaters	0.05–0.10	0.12–0.23
Sluggish streams and large lakes	0.10–0.15	0.23–0.35
Large streams of low velocity	0.15–0.20	0.35–0.46
Large streams of normal velocity	0.20–0.30	0.46–0.69
Swift streams	0.30–0.50	0.69–1.15
Rapids and waterfalls	>0.50	>1.15

noted that the waste effluent to a stream may be present as a point source or a nonpoint source. A *point source* is defined as a source where the pollutant enters the stream at a specific place, such as the effluent pipe from a sewage treatment plant. An example of a *nonpoint source* would be drainage from a stockyard or farming area where waste enters the stream over a long section of the stream bank. In the model used here, both of these source terms are simplified by assuming a well-mixed stream. This simplification is possible because, for example, if the effluent pipe from a sewage treatment plant releases treated wastewater containing 5% of the original BOD content of the raw sewage into the middle of a stream, after the water has traveled a few meters down the channel, water at each side of the bank will still be clean, whereas water in the middle of the channel will start to experience lower oxygen levels, due to microbial degradation of the introduced waste. However, after a short amount of time (or distance downstream), most streams will be completely mixed and the BOD concentration will be uniform throughout the stream cross section. When this situation develops, the general equation (28-3) can be used. A similar argument can be made for nonpoint sources and stream mixing.

The average water velocity is represented by v. This value is easily measured and is usually given in the problem statement. The initial oxygen deficit (D_0) is calculated by subtracting from the saturation value the dissolved oxygen in the stream immediately downstream from the input. The value plotted in Fate is a result of subtracting the stream DO concentration above the waste input ($x < 0$) from the oxygen deficit calculated from the governing equation. The net result is $D_0 - D$, which is the remaining DO concentration in the stream.

The dissolved oxygen sag curve can be divided into several zones delineated by the dissolved oxygen concentration and the presence of specific biological communities. Each of these is shown in Figure 28-2. Above the point of waste entry, a *clean water zone* [labeled (1) in Figure 28-2] is present and is usually characterized by clear, fresh water containing a stable and natural fish, macro-invertebrate, and plankton population. DO levels are usually near saturation. As the wastewater enters the stream, a short *zone of degradation* is established [labeled (2) in Figure 28-2]. The water is usually more turbid and sunlight is reduced with depth in the stream. Chemical characteristics include (1) up to a 40% reduction of DO from the initial value, an increase in CO_2, and nitrogen present in organic forms. Biologically, bacterial activity increases, green and blue-green algae are present, fungi appear, protozoa (ciliates) are abundant, tubiflex and bloodworms are present, and large plants may die off.

The zone of active decomposition [labeled (3) in Figure 28-2] followes the zone of degradation. Physical characteristics of this zone include water that is gray or black in color, the presence of offensive odors, and no light penetration through the water. As the water travels through this zone, the DO concentration starts at 40% of the initial value, may drop to 0, and eventually returns to 40% of the initial value. Gases such as H_2S, CH_4, and NH_3 are usually produced by reducing conditions and contribute to the offensive odor. As O_2 levels drop, bacteria and algae may be the only life-forms present in the water column.

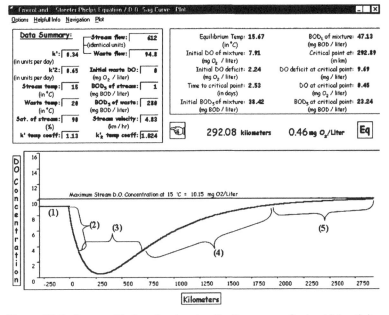

Figure 28-2. Streeter–Phelps plot showing the five zones of microbial activity.

A relatively long *zone of recovery* [labeled (4) in Figure 28-2] follows and is characterized by clearer water than that in the two preceding zones. Chemical characteristics include DO concentrations from 40% of the initial value up to saturation, decreasing CO_2 levels, and nitrogen present as NH_3 and organic forms. Biological characteristics include decreased numbers of bacteria and the presence of protozoa, bluegreen, green algae, tubiflex, and bloodworms. A *zone of cleaner water* [labeled (5) in Figure 28-2] is reached when the physical, chemical, and biological characteristics of the stream have nearly returned to the conditions present upstream of the pollution source.

With respect to these zones, one point of special interest is that at which the DO concentration (D) reaches its minimum value, referred to as the *critical dissolved oxygen concentration* (D_c). This point can be characterized by (1) the time required to reach this point (the critical time, t_c) and/or by (2) its distance downstream from the point source (the critical distance, x_c).

The time required to reach the critical distance can be calculated by

$$t_c = \frac{1}{k_2' - k'} \ln \frac{k_2'}{k'} \left[1 - \frac{D_0(k_2' - k')}{k' \cdot \text{BOD}_L} \right] \tag{28-4}$$

where D_0 is the oxygen deficit (O_2 saturation value − mixture value). The critical distance is calculated by

$$x_c = vt_c \tag{28-5}$$

where the water velocity, v, can be given in miles or kilometers. The critical dissolved oxygen concentration (D_c) can be calculated by

$$D_c = \frac{k'}{k_2'} \text{BOD}_L \cdot e^{-k'(x_c/v)}$$

REFERENCES

Metcalf & Eddy, Inc., *Wastewater Engineering: Collection, Treatment, Disposal*, McGraw-Hill, New York, 1972.

Sanitary District of Chicago, *Report of the Engineering Board of Review*, part III, Appendix I, 1925.

Till, J. E. and Meyer, H. R. (eds.), *Radiological Assessment: A Textbook on Environmental Dose Analysis*, NUREG/CR-3332, ORNL-5968, U.S. Nuclear Regulatory Commission, Washington, DC, Sept. 1993.

ASSIGNMENT

1. Install Fate on your computer (Fate is included with your lab manual). Open the program and select the river step, then the Streeter–Phelps module. A sample data set will load automatically. Work through the example problem, referring to the background information given earlier and the explanation of the example problem (included in Fate) as needed.

2. Construct a pollution scenario for your simulations. This will require you input data on a specific stream, such as flow rate, water temperature, background BOD concentration, and the most appropriate reaeration rate (values are given in the table of reaeration rates included in Fate and in Table 28-1). You will also need information for a wastewater treatment plant (flow rate, water temperature, k_2', BOD_L, etc.). For your initial simulation, assume that the wastewater enters the stream directly, without treatment.

3. Perform a simulation using your basic input data and evaluate the effluent DO concentration downstream. Next, perform a sensitivity test by selecting several input variables, such as mass loading, flow rates (to reflect an unusually wet or dry season), and first-order rate constants (those given in the table are only estimates).

4. Next, imagine that a wastewater treatment plant has been installed removing 95% of the BOD in your influent sewage. Change the input parameters accordingly and evaluate the effectiveness of your treatment plant in protecting the stream. Next, determine the percent removal of the influent sewage necessary to avoid the presence of a zone of active decomposition downgradient from your treatment plant.

5. Write a three- to five-page paper discussing the results of your simulations. Include tables of data and/or printouts of figures from Fate. A copy of your report should be included in your lab manual.

To Print a Graph from Fate

For a PC

- Select the printable version of your plot (lower right portion of the screen).
- Place the cursor over the plot at the desired x and y coordinates.
- Hold the alt key down and press print screen.
- Open your print or photoshop program.
- Paste the Fate graph in your program by holding down the control key and press the letter v.
- Save or print the file as usual.

For a Mac

- Select the printable version of your plot.
- Hold down the shift and open apple key and press the number 4. This will place a cross-hair symbol on your screen. Position the cross-hair symbol in the upper right corner of your plot, click the cursor and drag the cross-hair symbol over the area to be printed or saved, release the cursor when you have selected the complete image. A file will appear on your desktop as picture 1.
- Open the file with preview or any image processing file and print it as usual.

APPENDIX A

PERIODIC TABLE

Periodic Table of the Elements

Legend:

Element	Carbon
Element #	6
Symbol	C
Atomic Mass	12.01

IA	IIA	IIIB	IVB	VB	VIB	VIIB	VIII	VIII	VIII	IB	IIB	IIIA	IVA	VA	VIA	VIIA	VIIIA
Hydrogen 1 **H** 1.01																	Helium 2 **He** 4.00
Lithium 3 **Li** 6.94	Beryllium 4 **Be** 9.01											Boron 5 **B** 10.81	Carbon 6 **C** 12.01	Nitrogen 7 **N** 14.01	Oxygen 8 **O** 16.00	Fluorine 9 **F** 19.00	Neon 10 **Ne** 20.18
Sodium 11 **Na** 22.99	Magnesium 12 **Mg** 24.31											Aluminum 13 **Al** 26.98	Silicon 14 **Si** 28.09	Phosphorous 15 **P** 30.97	Sulfur 16 **S** 32.07	Chlorine 17 **Cl** 35.45	Argon 18 **Ar** 39.95
Potassium 19 **K** 39.10	Calcium 20 **Ca** 40.08	Scandium 21 **Sc** 44.96	Titanium 22 **Ti** 47.88	Vanadium 23 **V** 50.94	Chromium 24 **Cr** 52.00	Manganese 25 **Mn** 54.94	Iron 26 **Fe** 55.85	Cobalt 27 **Co** 58.93	Nickel 28 **Ni** 58.69	Copper 29 **Cu** 63.55	Zinc 30 **Zn** 65.39	Gallium 31 **Ga** 69.72	Germanium 32 **Ge** 72.61	Arsenic 33 **As** 74.92	Selenium 34 **Se** 78.96	Bromine 35 **Br** 79.90	Krypton 36 **Kr** 83.80
Rubidium 37 **Rb** 85.47	Strontium 38 **Sr** 87.62	Yttrium 39 **Y** 88.91	Zirconium 40 **Zr** 91.22	Niobium 41 **Nb** 92.91	Molybdenum 42 **Mo** 95.94	Technetium 43 **Tc** 99	Ruthenium 44 **Ru** 101.07	Rhodium 45 **Rh** 102.91	Palladium 46 **Pd** 106.42	Silver 47 **Ag** 107.87	Cadmium 48 **Cd** 112.41	Indium 49 **In** 114.82	Tin 50 **Sn** 118.71	Antimony 51 **Sb** 121.75	Tellurium 52 **Te** 127.60	Iodine 53 **I** 126.90	Xenon 54 **Xe** 131.29
Cesium 55 **Cs** 132.91	Barium 56 **Ba** 137.33	Lanthanum 57 **La** 138.91	Hafnium 72 **Hf** 178.49	Tantalum 73 **Ta** 180.95	Tungsten 74 **W** 183.85	Rhenium 75 **Re** 186.21	Osmium 76 **Os** 190.2	Iridium 77 **Ir** 192.22	Platinum 78 **Pt** 195.08	Gold 79 **Au** 196.97	Mercury 80 **Hg** 200.59	Thallium 81 **Tl** 204.38	Lead 82 **Pb** 207.2	Bismuth 83 **Bi** 208.98	Polonium 84 **Po** 209	Astatine 85 **At** 210	Radon 86 **Rn** 222
Francium 87 **Fr** 223	Radium 88 **Ra** 226	Actinium 89 **Ac** 227	Rutherfordium 104 **Rf** 261	Dubnium 105 **Db** 262	Seaborgium 106 **Sg** 263	Bohrium 107 **Bh** 262	Hassium 108 **Hs** 265	Meitnerium 109 **Mt** 266	Ununnilium 110 **Uun** 269	Unununium 111 **Uuu** 272	Ununbium 112 **Uub** 277						

Lanthanide Series

Cerium 58 **Ce** 140.12	Praseodymium 59 **Pr** 140.91	Neodymium 60 **Nd** 144.24	Promethium 61 **Pm** 147	Samarium 62 **Sm** 150.36	Europium 63 **Eu** 151.97	Gadolinium 64 **Gd** 157.25	Terbium 65 **Tb** 158.93	Dysprosium 66 **Dy** 162.50	Holmium 67 **Ho** 164.93	Erbium 68 **Er** 167.26	Thulium 69 **Tm** 168.93	Ytterbium 70 **Yb** 173.04	Lutetium 71 **Lu** 174.97

Actinide Series

Thorium 90 **Th** 232.04	Protactinium 91 **Pa** 231	Uranium 92 **U** 238.03	Neptunium 93 **Np** 237	Plutonium 94 **Pu** 244	Americium 95 **Am** 243	Curium 96 **Cm** 247	Berkelium 97 **Bk** 247	Californium 98 **Cf** 251	Einsteinium 99 **Es** 252	Fermium 100 **Fm** 257	Mendelevium 101 **Md** 258	Nobelium 102 **No** 259	Lawrencium 103 **Lr** 260

INDEX

Active laboratory notebook, 4
Alkalinity, 245, 246, 251, 253

Beer's law, 102
Biochemical oxygen demand (BOD), 217, 220–223, 227, 228, 317, 320, 321

Capillary column GC, 33, 46, 63, 64, 66, 69, 88, 113, 115, 117, 170, 171, 173, 186
Carbon dioxide (CO_2), 33, 51, 53–55, 58, 247, 248, 249
CFC, 58
Chlorinated pesticides, 39, 42, 83, 84, 86
Chromophores, 103
Coefficient of regression, 10

DDT, 39, 43, 83, 92, 152, 189
Detection limit, 8, 18
Diffusion, 280
Dispersion, 293, 296, 305, 306, 308, 309, 312
Dissolved oxygen (DO), 207, 209, 212, 217, 219–221, 318
Distribution coefficient (K_d), 191, 193, 196–199, 297

EDTA, 151, 162, 259–262
Electroneutrality, 74, 82

Fate and transport, 277, 285, 293, 303

Flame atomic absorption spectroscopy (FAAS), 73, 78–80, 127, 129, 131, 151–153, 158–161, 191, 195, 201

Gasoline, 61, 62, 64, 113, 114, 117
Global warming, 49, 52
Greenhouse effect, 49
Groundwater sampling, 25

Hardness, 257
Henry's law constant, 33–36, 45
High performance liquid chromatography (HPLC), 115, 143–145, 167, 170, 171, 173

Inactive laboratory notebook, 4
Inductively coupled plasma (ICP), 164
Infrared (IR), 49, 51, 52, 56, 58
Internal standard, 42, 86, 90, 179, 183
Ion chromatograph (IC), 73, 76–79
Ion-specific electrodes, 93, 151, 163

Limit of linearity, 102
Limit of quantitation, 102
Linear least squares analysis, 8, 148

Mass balance, 233

Natural organic matter (NOM), 84, 168, 172
Nitroaromatics, 143

Environmental Laboratory Exercises for Instrumental Analysis and Environmental Chemistry
By Frank M. Dunnivant
ISBN 0-471-48856-9   Copyright © 2004 John Wiley & Sons, Inc.

pC-pH, 252, 267, 268, 275
Polychlorinated biphenyls (PCBs), 39, 83, 86, 152
Precipitation, 123, 130–132
Propagation of uncertainty (POU), 10, 13, 17

Releasing agent, 154, 159–161

Sediment sampling, 25
Signal-to-noise ratio, 104, 107
Soil sampling, 26, 27
Soxhlet, 179, 181, 184
Standard addition, 152
Standard analysis plan, 19
Standard deviation, 9, 13, 15, 16
Standard operation procedure, 19

Statistical analysis, 7
Student's *t* test, 7, 10, 17, 91, 108

Total dissolved solids (TDS), 234, 239
Total solids (TS), 237
Total suspended solids (TSS), 233, 238
Tenax, 34, 39, 41, 42

UV-Visible, 101, 102

Vostok ice core, 53, 54

Water sampling, 22, 24, 30
Winkler titration, 207, 210, 211, 229
Working laboratory notebook, 4

Printed and bound by CPI Group (UK) Ltd, Croydon, CR0 4YY

16/04/2025

14658356-0001